Environmental Pollution

Atmosphere, Land, Water, and Noise

H. M. Dix
Department of Science,
Huddersfield Technical College

JOHN WILEY & SONS

Chichester · New York · Brisbane · Toronto

British Library Cataloguing in Publication Data:

Dix, H. M.
  Environmental pollution.—(Institution of
  Environmental Sciences. Series).
  1. Pollution
  I. Title   II. Series
  301.31   TDI74   80–40287

  ISBN 0 471 27797 5 (Cloth)
  ISBN 0 471 27905 6 (Paper)

Photosetting by Thomson Press (India) Limited, New Delhi
and printed in the United States of America by Vail-Ballon Press,
Inc., at Binghamton, N.Y.

Contents

Introduction to Series

In the last two decades the environment has figured increasingly in our everyday discussions. From a matter of superficial interest to all but an enlightened few, it has become not just the calling of professors and politicians, but the concern of parliaments and the responsibility of us all. This explosive increase of interest has been associated with an ever-widening accumulation of related environmental facts. This series seeks to bring together these facts, and allied ideas, so that they may be readily accessible in a number of identifiable volumes. The Institution of Environmental Sciences in promoting the series has, as an objective, the express desire to diffuse a balanced professional view on all matters relating to the environment.

The environment now figures prominently as a curriculum subject and Environmental Science is established as a school and university examination subject at all levels. The need for a comprehensive series of volumes which disseminate the wealth of information relating to this science has become of paramount importance. The series should be of value both to students of the environment and to professionals working in related fields.

PETER FARAGO
DAVID HUGHES-EVANS
JOHN F. POTTER

Preface

We are all aware today that pollution is a problem. There is a steady output of information and discussion in the daily press, in periodicals, and by the sound and vision media. Some of this output is highly emotive and is produced either by prejudiced individuals and pressure groups, or as reports on topical pollution incidents. It has been stated that pollution is caused by the deliberate or accidental contamination of the environment with man's waste, and a continuation of this practice will eventually cause the whole planet to become uninhabitable. Alternative views are that there has been pollution of the environment for a long time, and so far this has not proved to be harmful to the majority of the human population. Provided that critical and isolated incidents of pollution are dealt with as they occur, then there is no need to be concerned about the future. The effects and the extent of environmental pollution, both now and in the future, are a matter of doubt and controversy for many people.

This book does not take an extreme view, but recognizes that pollution is an increasing environmental problem that needs serious objective study. This requires factual information relating to waste production and disposal, the levels of pollution, and the effects that actually exist in our environment. Against this factual background, the present methods and implementation of pollution control can then be assessed. Pollution problems must be considered also within the context of our modern technological society and government economic policies. A reduction of pollution can only be achieved by adequate pollution control, and an objective appreciation of the problems and consequences by government, industry, and the consumer. The effectiveness of pollution control and the determination of government and industry to provide the necessary resources for this are crucial factors in the future development of environmental pollution.

Pollution is an integrated occurrence that must be considered within the natural systems that exist in the environment. However, in order to study pollution it is usual to subdivide it into the artificial divisions of atmosphere, land, water, and noise pollution. Pollution problems involve an understanding of some basic knowledge of biology, chemistry, and physical environmental phenomena. Consequently sufficient scientific background has been included in the book to enable the reader to understand the technicalities of pollution.

The information and views presented are intended to provide a basic introduction to environmental pollution. They are intended to inform and promote discussion for teachers and students who take environmental studies up to the level of a first-degree course. They will also prove useful to people and organizations who are concerned about the deterioration of the quality of the environment in which they live and work.

The author wishes to express his appreciation of the help and encouragement given by David Hughes-Evans and Dr Peter Farago, and his thanks to Dr Howard A. Jones of John Wiley and Sons Ltd for his help during the passage of the book through the press.

November 1979

SECTION I

General

CHAPTER 1

Introduction

1.1 The Problem

The existence of pollution in the environment, as a national and a world problem, was not generally recognized until the 1960s. Today many people regard pollution as a problem that will not go away, but one that could get worse in the future. It is increasingly being appreciated that the general effects of pollution produce a deterioration of the quality of the environment. This usually means that pollution is responsible for dirty streams, rivers and sea shores, atmospheric contamination, the despoilation of the countryside, urban dereliction, adulterated food, etc. All these deleterious factors are affecting the environment in which people reside, work, and spend their leisure time.

The present increasing emphasis upon pollution may create the impression that there has been a relatively sudden deterioration of the environment, that was not apparent twenty or thirty years ago. This is not the case. Pollution must have started at the time when man began to use the natural resources of the environment for his own benefit. As he began to develop a settled life in small communities, the activities of clearing trees, building shelters, cultivating crops, and preparing and cooking food must have altered the natural environment. Later, as the human population increased and became concentrated into larger communities which developed craft skills, there were increasing quantities of human and animal waste and rubbish to be disposed of. In the early days of man's existence the amount of waste was small. It was disposed of locally and had virtually no effect upon the environment. Later, when larger human settlements and towns were established, waste disposal began to cause obvious pollution of streets and water courses. In the thirteenth century the prevalence of cholera, typhus, typhoid and bubonic plague was associated with the lack of proper waste disposal methods. By the mid-nineteenth century the population of the UK had increased to 22 million, and many canals and rivers were grossly polluted with sewage and industrial waste. Some sewerage systems existed in towns, but the collected sewage was discharged into the nearest river without any treatment. Salmon had completely disappeared from the River Thames and outbreaks of cholera still occurred in London. A Royal Commission on the Prevention of River Pollution was established in 1857, and eventually the first preventive river pollution legislation was passed in 1876 and 1890. However, there was little significant improvement in pollution until after the First World War, and the

3

condition of rivers had deteriorated again by the end of the Second World War. Even today, a number of British and continental coastal towns discharge almost untreated sewage into near-shore waters.

The increasing pollution of land and water was accompanied by air pollution. This must have begun as soon as man started to use wood fires to provide 'space heating' and a means of cooking food. Later, surface soft coal was discovered and used as a fuel, and records show that coal smoke was a nuisance in London in the thirteenth century. In 1273, Edward I made the first ever anti-pollution law to prevent the use of coal for domestic heating, so smoke pollution has been recognized for at least 700 years. However, smoke pollution in London continued and is recorded in both the sixteenth and seventeenth centuries. In the late eighteenth and throughout the nineteenth centuries there was a marked increase in air pollution, because of the greater use of coal by developing industry. From 1750, the chemical industry began to develop, and this caused the discharge of acid fumes into the smoky air of some manufacturing towns. A Royal Commission was set up in 1862 to consider air pollution and this resulted in the first Alkali Act in 1863, which set limits to the concentration of acid in discharged waste gases. However, the increasing domestic and industrial combustion of coal, and the production of piped coal gas from 1815, caused air pollution to steadily get worse. Large cities were particularly affected, and the well known 5 day smog incident in London in 1952 directly contributed to the deaths of 4000 people (Figure 6.1). As a result, the Beaver Committee on Air Pollution was established in 1953, and the Clean Air Act was passed in 1956. This was the first effective statute to provide the means of controlling atmospheric pollution.

Noise pollution probably started when man first developed machines. The increase in industrial plants in the nineteenth century produced indoor noise pollution of the working environment for many factory and mill workers over a 6 day week. Outdoors, the development of private and public transport in Britain brought environmental noise, as the railway services came into use during the 1830s, motor transport from 1900, and regular aeroplane services from 1922. During the first half of the twentieth century environmental noise considerably increased, but it was not recognized as pollution. Industrial and outdoor noise was designated as 'nuisance' when the Noise Abatement Act was passed in 1960. Whereas the earlier increase in noise occurred in work places and in connection with transport, during the last thirty years noise has spread into the home and places of leisure and entertainment. Many domestic labour-saving appliances are very noisy, and it has become commonplace to have a semi-continuous background of music for many human activities throughout the day. The type of noise and the sound intensities that are tolerable to the ear, or are not pollutive, are controversial, because the human reaction to noise is largely a matter of personal opinion and choice. However, it is accepted that any noise that affects the human hearing range and response must be regarded as a form of pollution.

Certainly the most rapid increase in environmental pollution has taken place during the last 150 years, and it has been attributed to a number of interrelating factors. In 1850, the population of the UK was about 22 million, but by 1920 it had doubled to 44M people over a 70 year period. During the next 56 years, by 1976, there was a further 27% increase to about 56M people. This exponential type of population increase was accompanied by the need for more houses, increased production of manufactured and consumer goods, increased numbers of road vehicles, and increased consumption of fossil fuels and resources. All this increased manufacturing output and energy consumption has resulted in more and more waste which has to be disposed of into the environment. The industrial revolution that mainly occurred in the nineteenth century began the development of modern technology. Many new and increasingly complex manufacturing processes and plants were developed, and these produced increasing amounts of gaseous, solid, and liquid wastes into the environment. At the same time the complementary extraction of mineral ores and coal resources caused de-spoilation of the countryside, and created large quantities of mining spoil and land pollution. The industrial expansion of the nineteenth century has continued in this century, which has been notable for the rapid development of many new technologies. These not only produce waste, but the increasing diversity of content and potential toxicity increase the problems of safe waste disposal within the environment. Accompanying new technology there is always obsolescence. Many of the technical processes started in the last century are no longer being operated, and so obsolete industrial machinery and buildings, houses, railway lines, and canals become abandoned and derelict, causing an increase in land pollution.

The political and economic policies carried out by successive British governments have also affected pollution problems. These policies have been based upon national economic growth, and governments have introduced measures to improve the standard of living of the population. All sectors of industry have been encouraged and exhorted to expand their technologies and productivity. One result of this policy has been the increasingly rapid obsolescence of processes, equipment, and consumer goods. At the same time monetary inflation within the national economy and rapid rises of wages and salaries enables consumers to buy new products and rapidly replace old ones. The increased production, and the rapid turnover of manufactured goods, has significantly contributed to the production of more manufacturing waste and discarded consumer rubbish to be disposed of in the environment.

The developments described have not been solely responsible for the increased pollution. They caused more waste to be produced, but most pollution is caused by the methods used for its disposal within the environment. Disposal is carried out by all sections of the population, which include private and public industry, Local Authorities, and individual people. Their attitude towards waste disposal is of considerable importance in relation to the pollution effects that are caused. Pollution is often regarded as the problem of neither industry nor the individual, but the devolved responsibility of Local Authorities and Central Government. Pollution in industrial areas has long been accepted as inevitable, and part of their environment. This attitude has been typified by the northern England expression 'Where there's muck there's brass', showing the association of pollution and economic gain. Control and prevention of pollution by the responsible disposal of waste is expensive, and the money required is considered by many industrial firms as a non-productive cost, to be reduced to the minimum. Government policies prior to the 1970s did not greatly help to control or reduce pollution, because of the low priority given to legislation and finance. Often action only appeared to be taken as a result of public concern and the pressure of isolated pollution incidents, for example the Clean Air Act 1956, and the Deposit of Poisonous Wastes Act 1972.

However, during the 1970s the national attitude towards pollution has been changing. Successive governments have passed a number of statutory Acts to provide the means of controlling pollution. The whole subject is being considered with a more integrated approach. This is shown by the Control of Pollution Act 1974, which for the first time in one statute covers the areas of atmosphere, land, water, and noise pollution. Also the reorganization of the public water and sewage undertakings in 1974 has produced regional responsibility by single authorities for water supply, waste treatment, and water pollution prevention. Industrial firms are having to adopt a more responsible attitude to their waste treatment and disposal, as a result of new legislation and rising charges for waste treatment and disposal imposed by the Water Authorities. Lastly the continuing discussion of pollution problems by the press, broadcasting and numerous environmental pressure groups is attempting to make the general public more aware of the seriousness of pollution in their environment.

1.2 Pollution in General Terms

Pollution can be defined in various ways. One type of definition involves the use of subjective judgements, for example the report *Pollution: Nuisance or Nemesis;* (1972), describes pollution as 'the deliberate or accidental contamination of the environment with man's waste'. Alternatively, pollution has been expressed as 'matter in the wrong place', or 'anything released into the environment which degrades it', or 'the presence of matter or energy in an unusual or unintended place'. Other definitions are more objective and scientific. T. J. McLoughlin defines pollution as 'the introduction by man of waste matter or surplus energy into the environment, which directly or indirectly causes damage to man, and his environment other than himself, his household, those in his employment, and those with whom he has a direct trading relationship'. This statement itemizes the areas where pollution damage can occur, but is entirely concerned with man rather than the whole biosphere. Holister and Porteous, in their *Dictionary of the Environment*, give a very comprehensive definition as follows. 'A pollutant is a substance or effect which adversely alters the environment by changing the growth rate of species, interferes with the food chain, is toxic, or interferes with health, comfort, amenities, or property values of people'. They add that 'generally pollutants are introduced into the environment in significant amounts in the form of sewage, waste, accidental discharge, or as a by-product of a manufacturing process or other human activity. A polluting effect is normally some kind of waste energy such as heat, noise or vibration'. In very general terms pollution causes degradation and/or damage to the natural functioning of the biosphere. The damage caused may be briefly summarized as follows:

(1) Damage to human health caused by specific chemical substances present in the air, food and water, and radioactivity.

(2) Damage to the natural environment which affects vegetation, animals, crops, soil, and water.

(3) Damage to the aesthetic quality of the environment caused by smoke, chemical fumes, dust, noise, the dumping of waste and rubbish, and dereliction.

(4) Damage caused by long-term pollution effects which are not immediately apparent. The insidious effects are caused by low-level pollution absorbed into the body over long periods of time, for example carcinogenic substances, radioactivity, and excessive noise.

CHAPTER 2

The Natural Environment

2.1 The Natural Cycles of the Environment

Man's environment in its widest sense is called the biosphere. This consists of the earth's crust, the surrounding atmosphere, and the various forms of life that exist in the zone 600 metres above and 10 000 metres below sea level. The biosphere is very large and complex and so is usually divided into smaller units or ecosystems. The concept of an ecosystem is defined as the plants, animals, and micro-organisms that live in a defined zone, and the physical factors present, for example soil, water, and air. Within an ecosystem, for example a river, there exist dynamic interrelationships between the living forms and their physical environment. These relationships can be expressed as natural cycles which provide a continuous circulation of the essential constituents necessary for life. The cycles mainly operate in a balanced state with little variation in the unpolluted natural environment. If this natural balance did not exist, then many life-sustaining processes would not be maintained, and the ecosystems would be variable and unstable. The balanced operation of natural cycles and ecosystems contributes to the stability of the whole biosphere, which is fundamental to the continued existence and development of life on earth.

Most ecosystems contain many different organisms, for example a deciduous forest can support over one hundred species of birds, as well as hundreds of species of shrubs, flowers, grasses, insects, and small mammals. Man's intervention by tree felling, land clearance, replanting of coniferous species, atmospheric pollution, stream pollution, deposition of poisonous wastes, etc. inevitably changes the pre-existing balanced ecosystem. Another example is man's drive to produce more food with greater efficiency through a monocrop agricultural system. This husbandry technique, where a single crop species is planted exclusively with no crop rotation, can result in soil mineral and humus deficiency, an increase of pests and plant diseases, and a reduction of natural predators. To offset these effects, the farmer may apply quantities of fertilizers to the soil, and spray the crop with fungicides and insecticides. This applied technology changes the natural and balanced ecosystems of the land concerned, and it can produce an ecological unbalance that needs constant adjustment by the application of chemicals that are potential pollutants. A third example can occur in an estuarine ecosystem which supports a balanced range of aquatic plant and animal life. The introduction of excessive fertilizer run-off from bordering agricultural land, heated cooling water from an electrical power station, and sewage effluent will tend to promote excessive plant growth or eutrophication, at the expense of the animal life. These three examples illustrate how man's activities can change natural ecosystems and these changes may be beneficial or disadvantageous to the environment. It is important to understand the operation of the various ecosystems when considering the short term and long term effects of pollution. The general effect of pollution within an ecosystem is to alter the prevailing balance of the natural cycles, and this is particularly relevant to certain cycles which operate in the atmosphere, on land, and in water.

The Hydrological Cycle

The hydrological cycle is a continuous natural process whereby water is exchanged between the atmosphere, the land, the sea, all living plants and animals, and industrial plants. It can be illustrated as in Figure 2.1. Water as rain, hail or snow is precipitated on to all land and water surfaces. Water on land

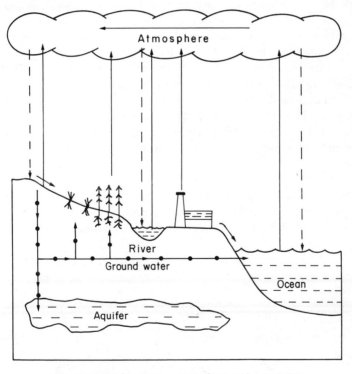

→ water evaporated as water vapour from soil, vegetation, surface waters, and oceans

--→ water vapour precipitated as rain, hail, or snow

—•→ water movement in the earth

Figure 2.1. The hydrological cycle.

surfaces eventually percolates into the soil as soil or ground water. Within the ground there is always a natural water level or water table. The soil below the water table level is saturated, and water is sustained by the underlying clay and rock strata. The depth of the water table is measured from the surface and depends upon the nature of the soil and underlying strata. Ground water does not remain static but moves in various directions. Water can move up above the water table by capillarity, and this provides a continuous supply of water to the surface layers of soil, where it is absorbed by plant roots, especially when there is no rain. Some ground water moves underground by filtering through the interstices of the soil or substratum in any direction. Water also emerges from the ground at lower altitude levels and flows into streams, rivers and lakes, and helps to provide man's water supplies. In some regions of the earth there are underground water-bearing layers of porous rock which are called aquifers. These are situated above impermeable rock strata, and water precolates through the porous rocks and forms underground lakes or reservoirs. Aquifers are another important water supply, and the water can be extracted by sinking wells or bore holes and pumping it to the surface.

Not all the water precipitated on land percolates into the soil, especially where the rainfall is high. Surface water or run-off flows into streams, rivers, lakes, and catchment storage areas or reservoirs. The ground surface and all water surfaces on the earth lose water by evaporation, using solar energy. Natural evaporation from the oceans exceeds precipitation by rain into the seas by about 9%. This 9% is eventually moved as water vapour over to the land surface, and so balances the hydrological cycle and provides additional water for man's needs. Some of the capillary ground water that is absorbed by plants passes out of the leaf surfaces as water vapour by the process of transpiration. This is an essential process which assists in ensuring conduction of water and dissolved mineral salts throughout the plant.

Thus the hydrological cycle consists of a balanced continuous process of evaporation, transpiration, precipitation, surface run-off, and ground water movements. The global cycle can be summarized as shown in Table 2.1. Each year an estimated 507 Tm^3 are evaporated, and the same quantity of water is

Table 2.1. Global volume of the hydrological cycle

Over land			Over sea		
	Daily	Annually		Daily	Annually
	(km^3)	(Tm^3)		(km^3)	(Tm^3)
Precipitated water	260	120		775	380
Evaporated water	160	80		875	420
Run-off	100	40	Difference	100	40

(1 m^3 equals 1000 litres; and 1 tera cubic metre (1 Tm^3) equals one million million cubic metres.)

precipitated over the whole surface of the earth. The amount of water which flows from the land to the sea in streams and rivers is 44.5 Tm^3 per year, and this is potentially available to supply man's water supplies.

The Nitrogen Cycle

The presence of nitrogen and its compounds in the biosphere is essential for the maintenance of life. Exchanges of nitrogen take place within ecosystems constituting the nitrogen cycle (Figure 2.2). Plants and animals continually produce proteins, which are organic compounds containing nitrogen. Plants absorb nitrates from the soil to produce proteins, and many animals feed on plants and obtain their proteins in this way. The death and excreta of organisms contribute organic residues containing proteins to the soil. In the soil there are different types of micro-organisms which utilize nitrogen compounds for their metabolism. Proteins are progressively acted upon by these bacteria, and a chain of intermediate compounds such as ammonia, nitrites, and finally nitrates is produced. Nitrates are absorbed by plants and so re-enter the nitrogen cycle. There is also an exchange of nitrogen between the cycle and the atmosphere, through the action of other soil micro-organisms. Some break down soil nitrate into nitrogen by a denitrification process, and others convert nitrogen into soluble nitrogen compounds in the soil at a rate of about 0.097–0.105 tonnes per hectare per year. In the total cycle about 3.7–7.4 tonnes of nitrogen per hectare is added to the soil each year. Against this, there is a loss from the soil through the leaching of nitrates into fresh water courses and the sea. The natural nitrogen

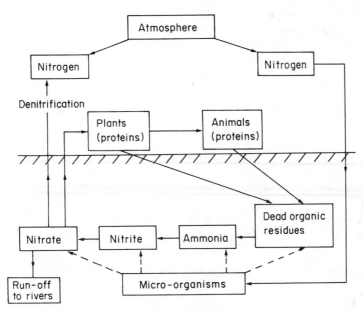

Figure 2.2. The nitrogen cycle.

cycle is balanced, and so the concentration of nitrogen in the atmosphere remains relatively constant. Modern agricultural husbandry involving very efficient crop production requires the addition of nitrogenous fertilizers to the soil. This adds about 40 M tonnes of nitrogen annually to the world's cultivated land. The result of this continuing practice is to upset the balance of the natural cycle. Pollution of water is taking place by the leaching of excessive amounts of nitrate from agricultural land into streams, rivers and lakes. In 1970, it was estimated that 8.5 M tonnes of leached nitrogen was discharged into the world's oceans, and today much more nitrogen fertilizer is used by farmers throughout the world to increase crop yields.

The Phosphate Cycle

Plants and animals require a continual supply of phosphate compounds. Organophosphates are essential for cell division involving the production of nuclear DNA and RNA, and phosphates are required for the growth and maintenance of animal bones and teeth. Plant and animal nutrition involves energy metabolic pathways with chemical reactions utilizing adenosine tri-phosphate (ATP). Land plants absorb inorganic phosphate salts from the soil, convert these into organic phosphate, and animals eat plants, so obtaining their phosphates. The basic source of phosphates is located in rocks and the soil, where phosphates exist in the soluble and insoluble or fixed form. Phosphate absorbed from the soil is returned to it in dead plant and animal organic residues, which are converted to humus by the action of soil micro-organisms. Much of the phosphate in the soil is fixed or adsorbed on to soil particles, but some is lost through leaching out into water courses (Figure 2.3). In fresh water the floating algae or phytoplankton rapidly absorb soluble inorganic phosphates and convert them into organophosphates. Algae provide food for zooplankton which in turn are consumed by other animals. All forms of life eventually die and their organic remains or debris settle to the bottom of the water. In due time, the organic debris decays through the action of micro-organisms, and phosphates

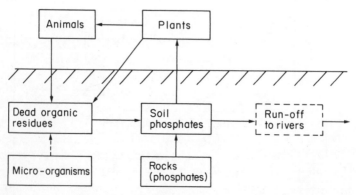

Figure 2.3. The phosphate cycle on land.

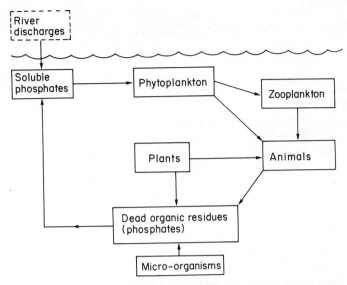

Figure 2.4 The phosphate cycle in water.

are released into the water for recycling again (Figure 2.4). The natural phosphate cycle can be substantially affected by pollution. Agricultural fertilizer applications containing superphosphate or triple superphosphate are now frequently used on the land; and sewage, even after treatment, contains phosphates derived from excreta and detergents. These phosphates can eventually reach fresh water streams and rivers through land run-off and effluent discharge. In the UK it has been estimated that 200 000 tonnes of phosphate enters sewers each year, and 132 000 tonnes or 66% of this is discharged into rivers. Phosphate pollution of rivers and lakes has caused excessive growth of algae, which depletes the water oxygen content and disrupts the natural food chains.

The Sulphur Cycle

Plants and animals require a continuous supply of sulphur and its compounds in order to synthesize some amino acids and proteins. Exchanges of sulphur take place within ecosystems through the activities of so-called sulphur bacteria. The circulation of sulphur and its compounds in the environment constitutes the sulphur cycle (Figure 2.5). The top half of the cycle shows the sulphur oxidation process. The lower sector illustrates the conversion of sulphate into plant and animal cellular proteins, and the degradation of dead plant and animal material by bacterial action. In very polluted waters there may be anaerobic conditions with no dissolved oxygen present, and the bacterial decay action then produces obnoxious hydrogen sulphide and deposits of black iron sulphide. In unpolluted or slightly polluted waters, aerobic or oxygenated conditions are present, and the

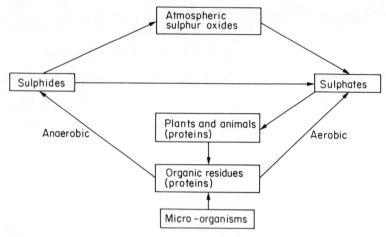

Figure 2.5. The sulphur cycle.

sulphur bacteria convert sulphides into sulphates for use in further protein production.

The Oxygen Cycle

Oxygen is a major component of all living organisms and an adequate supply is therefore vital for continued life in the biosphere. Oxygen is required by most plants and animals and all humans for aerobic respiration or the enzymic oxidation of organic food, which sustains growth and general metabolism. During daylight, plants carry out the biochemical process of photosynthesis, which produces carbohydrate and the liberation of oxygen. Therefore an oxygen cycle exists whereby the gas is absorbed from the environment during aerobic respiration, and released into the environment as a result of photosynthesis. There is also a continuous exchange of oxygen between the 20.9% in the atmosphere and all areas of water on the earth (Figure 2.6). The total amount of oxygen in the biosphere is relatively constant, and so the oxygen cycle is stable. Pollution effects may cause a deficiency of oxygen in localized aquatic situations.

2.2 Food Chains and Webs

Any ecosystem contains groups of organisms, or mini-populations of plants and animals which are interdependent in respect of food and energy. Organisms divide into three ecological classes in relation to their nutritional levels. Herbage such as grass is an example of the producer class, and this provides food for grazing animals which constitute the primary consumer class. Man is a secondary consumer when he eats primary consumers such as sheep and beef cattle. The third class are called decomposers, and use the dead remains of producers or consumers as their source of nutrition. The nutritional interdependence between

Atmosphere

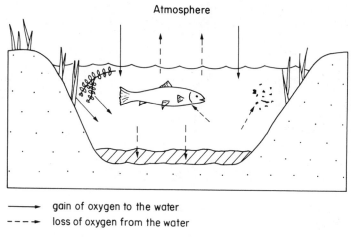

⟶ gain of oxygen to the water

– – – ➤ loss of oxygen from the water

Figure 2.6. Aquatic oxygen exchange.

different species of organism can be shown by the concept of a food chain. This is defined as a series of organisms through which energy is transferred. Each organism in the series feeds and derives energy from the proceeding one, as shown in Figure 2.7.

In a freshwater ecosystem there is a different food chain. Floating microscopic phytoplankton are an example of the producer class, which provide food for the primary consumers or zooplankton. These consist of microscopic animals and larval forms which in turn provide food for the larger secondary consumers such as fish. The tertiary consumers feed on fish and they may be predatory fish, or birds, or mammals, or anglers (see Figure 2.8). Producers derive their energy from photosynthesis, where solar energy is utilized for the combination of carbon dioxide and water to produce carbohydrate and oxygen. Primary consumers feed on producers and obtain their energy by the enzymic digestion of the plant species. Secondary and tertiary consumers are predatory animals which also derive their energy from the digestion of food. The example below shows other ecological features. The tertiary consumer animals are large in size and feed

Level or class	Type	Example
Producer	Herbage	Grass
↓	↓	↓
Primary consumer	Herbivores	Grasshopper
↓	↓	↓
Secondary consumer	Small carnivores	Thrush
↓	↓	↓
Tertiary consumer	Larger carnivores	Hawk

Figure 2.7 A simple linear food chain on land.

Level or class	Type	Example
Producer	Phytoplankton	Algae, diatoms
↓	↓	↓
Primary consumer	Zooplankton	Copepods, larvae
↓	↓	↓
Secondary consumer	Fish	Minnow
↓	↓	↓
Tertiary consumer	Bird	Kingfisher

Figure 2.8 A simple linear food chain in water.

on smaller animals at the secondary level. Also the tertiary consumers have relatively large energy requirements commensurate with their size, and so need to eat large numbers of the smaller secondary consumers. Considering the linear food chain overall and starting with the producers, then there are decreasing numbers of organisms at the different levels along the chain. This relationship is expressed in the concept of the pyramid of numbers, which shows the numerical relationship between organisms in the different ecological niches or local habitats (Figure 2.9). To this diagram should be added the decomposers, which consist of a large population of bacteria and fungi that exist in all types of ecosystems. They obtain their nourishment and energy from the organic chemical material present in the dead residues of all species at all levels of the pyramid. Decomposers are ecologically important because they remove dead organic material from the ecosystem, recycle it, and prevent it from accumulating. They have an important role in reducing the amount of pollutive material in water. Biochemically there are two chief types of decomposers. Aerobic types require oxygen for their nutrition, but anaerobic types can exist in low or no oxygen conditions.

The linear food chain described is a very simple example, but this type is the

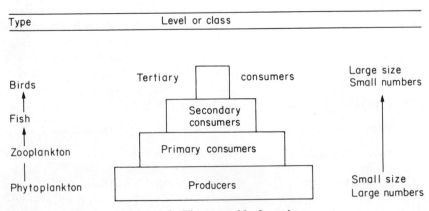

Figure 2.9. The pyramid of numbers.

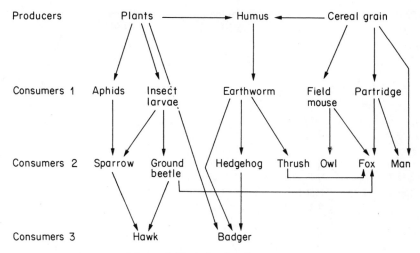

Figure 2.10 A food web on land.

exception rather than the rule in most ecosystems. The number of links in food chains vary, and there is often more than one species competing for food at each consumer level. If all the detailed nutritional relationships in any one ecosystem are worked out, the result is a series of interdependent food chains. This complex pattern is better described as a food web or food net, and one example is shown in Figure 2.10. This food web is a simplified representation of the more complex interrelationship which exists in a particular ecological niche. In a food web the basic operational principle is that each species is dependent upon at least one other species, and the numbers of each link species must be sufficient for their continued existence. Provided that these conditions are maintained, the food web will exist in an ecological nutritional equilibrium. However, food webs are vulnerable to changes in the environment, which may affect the food supply, reproductive rate, etc. There is a critical population density for each link species based upon the pyramid of numbers.

In an aquatic ecosystem a more complex food web will exist, as shown in Figure 2.11. This is also affected by changes in the environment such as variations in water temperature and oxygen content as well as nutrient supply, which can reduce or increase the population of link species. Both the food webs shown can be affected by pollution which causes changes in the ecological balance of the ecosystem. The presence of toxic chemical wastes and pesticides can reduce or completely eliminate the population of some link species and severely disrupt the food web. In fresh water, the presence of pollutive organic matter such as sewage or silage effluent reduces the dissolved oxygen content of the water. This in turn reduces or eliminates those link species which require aerobic conditions, and can promote an increase of the link species that do not require highly oxygenated water. Alternatively an increase of nutrients through run-off from agricultural land will stimulate plant producers to excessive growth,

18

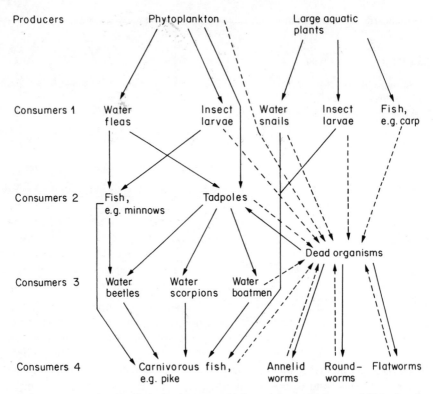

Figure 2.11. A food web in fresh water. (Adapted from *Biology—A Functional Approach*, 2nd edn, by M. B. V. Roberts (Nelson, 1974)).

with the result that the balance of the ecosystem is again upset. If the pollution is of short duration and not continuous, then the ecosystem will recover. This occurs when the link species and their numbers return to the pre-pollution state, and an integrated and balanced food web is re-established.

CHAPTER 3

Waste and Pollution

Most pollution is caused by the need to dispose of waste, which may be defined as any gaseous, solid, or liquid material that is discarded because it has no further apparent use for the owner, industrial processor or manufacturer. Waste cannot be eliminated but must be disposed of and contained within the global environment, and if waste products can produce harmful effects they are potential pollutants. Therefore when waste materials are released into the atmosphere, or dumped on land, or discharged into streams, rivers or the sea, they effectively pollute the environment.

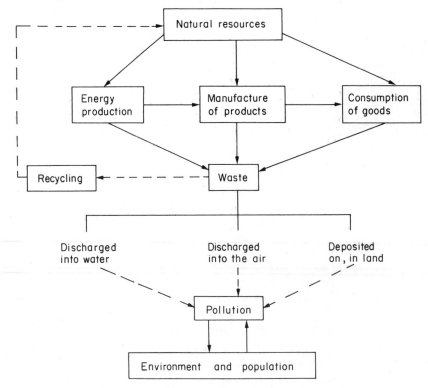

Figure 3.1. Waste and pollution within the manufacturing–consumer system.

19

All manufacturing processes and consumer activities produce waste and the relationship between productivity and consumption is shown in Figure 3.1. Not all waste needs to be disposed of into the environment as shown. Some waste can be recycled or returned to the production cycle, for example metals, glass, paper, rubber, textiles, and chemicals. Recycling reduces the amount of waste to be discarded and the need to use new natural resources for manufacturing processes. The above system is controlled by national policies and economics. If the present dictum is economic growth, then as production and the consumption of goods increases, the amount of waste produced increases. The problems of waste involve the quantities, composition and types of materials produced, and the methods of disposal within the environment.

In very general terms the sources, types and disposal methods of all waste materials are summarized in Figure 3.2. Gaseous and particulate matter are emitted into the atmosphere from all types of domestic, commercial and industrial premises. These wastes are an increasing source of atmospheric pollution and some material is eventually returned to the earth and deposited on to land and water surfaces. Solid wastes are mainly deposited on land, but the subsequent leaching out of chemical substances from the soil can be a source of water pollution. Liquid wastes may be discharged into water courses or directly into coastal waters, and so become a further source of water pollution. Therefore most wastes are deposited on, or buried in land, or discharged into water. The

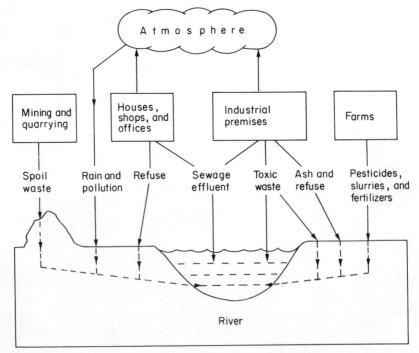

Figure 3.2. Sources and disposal of waste materials in the environment.

types of waste and the method of disposal are all interconnected within the biosphere, as shown in Figure 3.2. Whilst it may be convenient to consider land and water waste disposal separately, it is important to realize that the pollution caused is essentially an integrated effect.

The majority of wastes may be broadly classified into solids or liquids, or a mixture of these physical states. Liquids are often called effluents, which means a stream or liquid discharge usually produced from an industrial process or sewage works. There are various types of effluents, but basically they consist of a carrier liquid containing soluble or insoluble chemical substances. A slurry is a semi-liquid or thick liquid type of effluent containing considerable amounts of insoluble suspended solids. Solid waste products may be described according to particle size as rubble, gravel, grit, ash, or dust; or according to their origin as overburden or spoil from mining operations, or slag or clinker from industrial processes. The chemical constituents of wastes are loosely classified into organic or inorganic substance. The imprecise term 'organic' is used for compounds of carbon, other than its oxides, metallic carbonates and related compounds. Many organic waste compounds originate from living organisms, but others are manufactured. All other chemical compounds are described as non-organic or inorganic in composition. Small quantities of so-called biological wastes are produced by some food processing plants, and medical and research institutions. These solid and effluent wastes are characterized by the presence of micro-organisms and a wide range of complex organic compounds. A further type of solid and liquid waste is separately categorized because it contains radioactive material, and needs to be disposed of under special safety precautions in order to safeguard human health and the environment.

Wastes are produced from eight broad sectors, according to the type of producers or industry from which the solid and liquid waste originates.

1. Domestic sector. Domestic premises produce solid, and liquid wastes. These range from worn-out furniture, equipment and cars, gardening waste and dustbin refuse; to sewage containing excreta, grease, washing water, and detergents.

2. Commercial and Retail Trade sector. These premises produce mainly solids which are restricted to paper, board, and plastics, together with obsolete furniture and equipment.

3. Industrial Manufacturing sector. This waste is in the form of solids, liquid effluents, and slurries containing a range of organic and inorganic chemicals. Industrial processes are continually changing as new and modified technologies are developed. Consequently products, plant, and premises may become obsolete and worn out, so causing waste disposal and dereliction problems. Also many industrial processes use water for cooling purposes, and this can produce thermal pollution, if heated cooling water is released into streams, rivers and lakes.

4. Construction Industry sector. Mainly solid waste is produced consisting of brick, stone, mortar and cement rubble, wood, glass, metals, and plastics, as well as obsolete electrical and plumbing equipment and materials. This waste comes

SECTION II

Atmosphere

from four main types of operations, involving building new premises, adaption and modernization of existing premises, the demolition of buildings for land clearance or development, and new road construction.

5. Extractive Industry sector. This industry carries out various mining and quarrying operations involved in the extraction of coal, rock, slate, sand, metallic ores, and clay. The waste consists of unusable solid spoil material, and liquid slurry from the washing and grading of the extracted materials.

6. Agricultural sector. Organic wastes are produced from farms in the form of manure slurries, silage effluent, and dairy washings.

7. Food Processing Industry sector. This sector includes the production of meat and dairy products, deep-frozen and canned foods, and the processing of liquid and dried food derivatives, ranging from fruit preserves to flour, drinks, and beverages. Wastes range from unusable meat, vegetable and fruit material, to processing water containing organic chemicals such as fats, proteins, and pesticides.

8. Nuclear Industry and Power sector. This industrial sector produces cooling water; and solid, effluent and slurry wastes that are radioactive for periods of time ranging from a few days to thousands of years.

types of waste and the method of disposal are all interconnected within the biosphere, as shown in Figure 3.2. Whilst it may be convenient to consider land and water waste disposal separately, it is important to realize that the pollution caused is essentially an integrated effect.

The majority of wastes may be broadly classified into solids or liquids, or a mixture of these physical states. Liquids are often called effluents, which means a stream or liquid discharge usually produced from an industrial process or sewage works. There are various types of effluents, but basically they consist of a carrier liquid containing soluble or insoluble chemical substances. A slurry is a semi-liquid or thick liquid type of effluent containing considerable amounts of insoluble suspended solids. Solid waste products may be described according to particle size as rubble, gravel, grit, ash, or dust; or according to their origin as overburden or spoil from mining operations, or slag or clinker from industrial processes. The chemical constituents of wastes are loosely classified into organic or inorganic substance. The imprecise term 'organic' is used for compounds of carbon, other than its oxides, metallic carbonates and related compounds. Many organic waste compounds originate from living organisms, but others are manufactured. All other chemical compounds are described as non-organic or inorganic in composition. Small quantities of so-called biological wastes are produced by some food processing plants, and medical and research institutions. These solid and effluent wastes are characterized by the presence of micro-organisms and a wide range of complex organic compounds. A further type of solid and liquid waste is separately categorized because it contains radioactive material, and needs to be disposed of under special safety precautions in order to safeguard human health and the environment.

Wastes are produced from eight broad sectors, according to the type of producers or industry from which the solid and liquid waste originates.

1. Domestic sector. Domestic premises produce solid, and liquid wastes. These range from worn-out furniture, equipment and cars, gardening waste and dustbin refuse; to sewage containing excreta, grease, washing water, and detergents.

2. Commercial and Retail Trade sector. These premises produce mainly solids which are restricted to paper, board, and plastics, together with obsolete furniture and equipment.

3. Industrial Manufacturing sector. This waste is in the form of solids, liquid effluents, and slurries containing a range of organic and inorganic chemicals. Industrial processes are continually changing as new and modified technologies are developed. Consequently products, plant, and premises may become obsolete and worn out, so causing waste disposal and dereliction problems. Also many industrial processes use water for cooling purposes, and this can produce thermal pollution, if heated cooling water is released into streams, rivers and lakes.

4. Construction Industry sector. Mainly solid waste is produced consisting of brick, stone, mortar and cement rubble, wood, glass, metals, and plastics, as well as obsolete electrical and plumbing equipment and materials. This waste comes

CHAPTER 4

The Atmosphere

4.1 Composition of the Atmosphere

The atmosphere consists of a mixture of gases that completely surround the earth. It extends to an altitude of between 800 and 1000 km above the earth's surface, but is deeper at the equator and shallower at the poles. The gaseous components are unevenly distributed within the estimated mass of 5.14×10^{21} grams. About 99.9% of the mass occurs below 50 km and 0.0997% between 50 and 100 km altitude. The atmosphere is usually divided into four layers, as shown in Figure 4.1.

The troposphere or lower atmosphere is in contact with the earth's surface, and is about 8 km high at the poles and 16 km at the equator. In the unpolluted state its composition is as shown in Table 4.1. The troposphere also contains varying quantities of water of 1–4% by volume, with a maximum concentration at about 10–15 km altitude. The water may be present as a gas; or in condensed form as clouds, fog, or mist; or in solid form as ice, or snow.

The stratosphere layer is above the troposphere, and extend an average height of about 12 km to 50 km above the earth. It differs from the troposphere in that there is very little water vapour (3 ppm by mass) and no cloud formation present. Its gaseous composition is similar, but the gaseous mass is only 15% of the total atmosphere, and it contains significantly more ozone. Solar radiation, which passes through the atmosphere, contains short ultraviolet (UV) wavelengths. By means of photochemical reactions the UV radiation splits the atmospheric oxygen molecules (O_2) into atomic oxygen (O). These atoms are highly reactive and recombine to form triatomic molecules of ozone (O_3), which is an active oxidizing agent. Ozone exists in a so-called layer at a height of about 15–60 km in the upper stratosphere and mesosphere. The maximum concentration of 0.1–0.2 ppm occurs at 20–30 km above the earth. The ozone layer absorbs significant quantities of the short wave UV radiation that passes through it. This property gives it an important function in protecting the biosphere, and if the layer were reduced or not present, then the ionizing UV rays would destroy most or all of the life on earth.

The third atmospheric layer, or mesosphere, is considered to be at an altitude of about 50–80 km above the earth. It consists of less gaseous mass, no water vapour, and ozone is present throughout the layer.

The upper atmospheric layer is the thermosphere or heterosphere, which com-

Height (km)

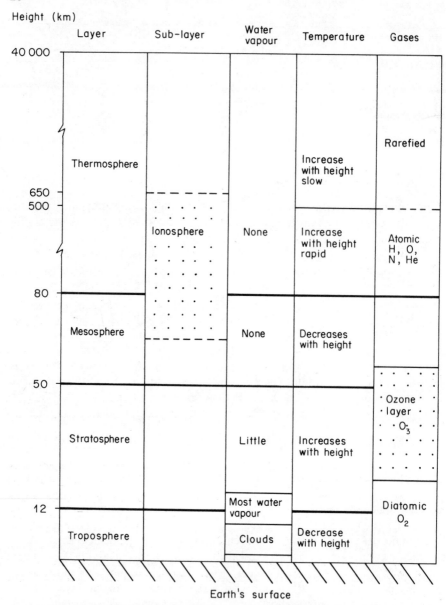

Figure 4.1. The composition of the atmosphere.

mences at an altitude of about 80 km. The gases present are in the atomic state and tend to seperate into layers. Between 80 and 115 km, oxygen and nitrogen are most abundant, but at 500 km, hydrogen and helium are the main constituents. Beyond 500 km the heterosphere becomes increasingly rarefied and extends to the extreme outer limit of the atmosphere at about 40 000 km. The layer contains no water vapour or ozone. Associated with the upper mesosphere and the lower

Table 4.1. Composition of the troposphere

Gas	% by volume of dry air	parts per million
Nitrogen	78.08	
Oxygen	20.95	
Argon	0.93	
Carbon dioxide	0.0325	325 (in 1974)
Neon	0.0018	18
Helium	0.0005	5.24
Krypton	0.0001	1.14
Ozone	0–0.00002	0–0.2

thermosphere is the ionosphere, so-called because it is a region where free electrons and ions exist. These are formed by the photochemical action of UV radiation upon nitrogen and oxygen gases.

4.2 The Troposphere

Atmospheric pollution is largely confined to this layer, with some penetration into the lower stratosphere. It is necessary to make a detailed consideration of the constituents of the layer to understand the effects of pollution.

Water

Water in the troposphere is continually being exchanged with the earth by a cyclic process of evaporation and precipitation. This constitutes the natural hydrological cycle (Figure 4.2) which has been described in Chapter 2 (Figure 2.1). In the troposphere, water vapour condenses to form clouds, rain, snow, or hail which is precipitated over the land and oceans, to the extent of about 507 Tm^3 per year (Table 4.2). This water may remain in the soil as ground water (10 080 Tm^3), or run off into water courses and eventually into the oceans (44.5 Tm^3). Also, wind and air movements transfer about 44.5 Tm^3 of water per year from the atmosphere over the oceans to the land masses. Thus in the unpolluted state, the natural hydrological cycle is balanced over an annual period. Overall the total water content of the troposphere is equivalent to a precipitation of 25.5 mm (1 inch) over the whole surface of the earth, or 2286 tonnes of rain per hectare.

The combustion of fossil fuels such as wood, coal, natural gas, and oil, the cooling towers of electrical power stations, and the exhaust emissions of transport vehicles all produce water vapour into the lower troposphere and increase its humidity. From about 2 to 10 km altitude the decreasing temperature of the air causes condensation of water vapour into clouds. These consist of small droplets of water surrounding condensation nuclei of about 0.1 micrometre (micron or μm) in diameter. Particulates in the troposphere serve as nuclei, and they may be soil particles, volcanic dust, sea salt crystals, chemical pollution

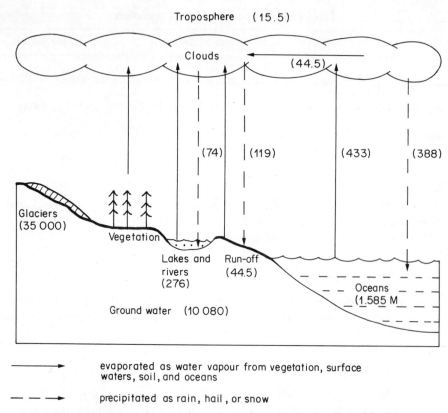

Figure 4.2. Exchanges of water within the hydrological cycle. Quantities in parentheses are water movements expressed in tera cubic metres (Tm³, or 10^{12} m³). (After *Earth Resources* by B. J. Skinner, Prentice Hall, 1969.)

particles, or those formed as a result of photochemical reactions. Below a height of about 2 km, lower night temperatures can cause condensation of water vapour into fogs and mists. Fogs consist of microscopic droplets of water with a diameter of 2–20 μm, suspended in the air. Mists differ only in the size of the droplets, which are less than 2 μm in diameter. Fogs are often associated with urban and industrial areas, where there are large volumes of water vapour emitted into the air from combustion processes, together with pollutants of particulate size.

Table 4.2. Recycled atmospheric water (Tm³ per year)

Evaporation		Precipitation	
From oceans	433	On to oceans	388
From land	74	On to land	119
Total	507	Total	507

Run-off from land to sea is 44.5 Tm³.
Transfer of atmospheric water from over the sea to land is 44.5 Tm³.

Carbon Dioxide

The amount of carbon dioxide (CO_2) in the troposphere is controlled by a cycle of gaseous exchange between the air and the biosphere, often called the carbon cycle (Figure 4.3). The diagram represents the land and marine biomass, which is the mass of living organisms that form a prescribed population in a given area of the earth's surface. The troposphere contains CO_2 (700×1000 M tonnes of carbon, or 700) and the soil contains a reservoir of CO_2 as organic compounds arising from dead organisms (700). The biomass contains 450×1000 M tonnes of carbon, and its living organisms evolve CO_2 into the troposphere as a result of respiration (10), and plants absorb CO_2 for the process of photosynthesis (35). Biomass organisms die and contribute carbon to the soil organic residues (25 per year). In the soil, living micro-organisms also respire and evolve CO_2 into the troposphere (25). In the total cycle, the CO_2 input and output to the soil balances at 25 per year, and in the troposphere the biomass exchange of CO_2 is also 25 per year. In the marine habitat, CO_2 is dissolved in the sea water and the biomass consists of phytoplankton (5) which provide food for fish (5). The phytoplankton

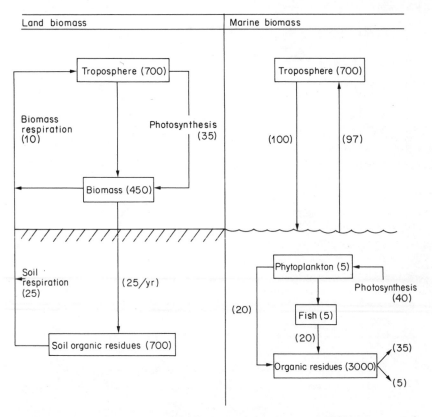

Figure 4.3. The carbon cycle (quantities in parentheses represent 1000 M tonnes of carbon as CO_2).

30

are plants which absorb CO_2 from the sea water for photosynthesis (40). The fish and phytoplankton die, and collectively contribute carbon $(20+20)$ to the reservoir of organic carbon residues (3000). These residues are slowly decomposed by micro-organisms to the extent of 40×1000 M tonnes of carbon per year. This process returns CO_2 to the surface waters (35), and to deeper waters (5). Within the marine carbon cycle, the CO_2 input of 40 is balanced by the

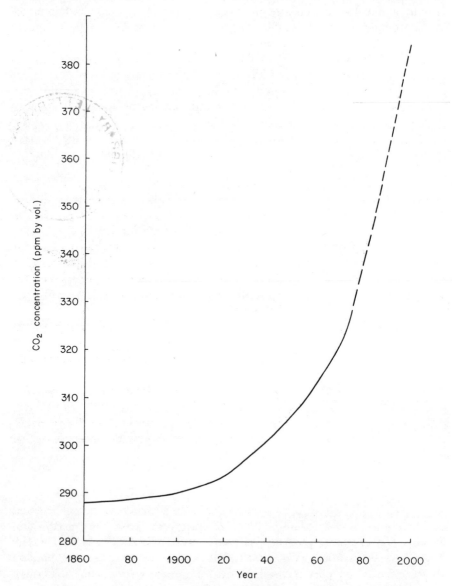

Figure 4.4. World atmospheric concentration of CO_2 from combustion of fuels, 1860–2000.

absorbed CO_2 of photosynthesis (40). There is also an exchange of CO_2 between the surface sea waters and the troposphere.

Considering the land and marine habitats together, then the total carbon cycle turnover is $35 + 40 \times 1000$ M tonnes of carbon per year. To this must be added 5000 M tonnes of carbon per year from the combustion processes of man's activity, so that the natural carbon cycle is not actually in balance. Measurements show that the concentration of CO_2 in the troposphere rose from 290 to 320 ppm (10%) over the period 1870 to 1970, and the increase between 1958 and 1970 was at an average rate of 0.2% per year (Figure 4.4). It was estimated in 1976 that the world combustion of about 5000 M tonnes of fossil fuels per annum was contributing the equivalent of 2.3 ppm of CO_2 to the atmosphere. The net annual increase of CO_2 in the air is 0.77 ppm, which indicates that 1.6 ppm of combustion CO_2 is being absorbed elsewhere, perhaps into the oceans and the biomass. A continued increase in world fossil fuel combustion in the future can be expected to produce a further increase of CO_2 in the troposphere, and this has been estimated at 20% by the year 2000.

Oxygen

Exchanges of oxygen take place between the biosphere and the troposphere. These constitute the oxygen cycle, which has been described in Chapter 2. The oxygen and carbon cycles are interlinked through the activities of living organisms. All vegetation in the biosphere takes in CO_2 and water to carry out the process of photosynthesis in light conditions. The respiratory process of plants and animals involves the utilization of oxygen to produce energy, and CO_2 and water are evolved. The oxygen absorbed for respiration is balanced by the output of oxygen from plant photosynthesis.

The combustion or oxidation of fossil fuels abstracts oxygen from the troposphere. It has been estimated that if all the known recoverable world reserves of fossil fuels were burned, the oxygen content of the troposphere would only be reduced by 0.15%, to 20.8% by volume. Similarly if there were no vegetation on the earth contributing photosynthetic oxygen, the effect upon the troposphere would be negligible. Therefore the oxygen content of the troposphere is relatively constant. The comparatively large reservoir of oxygen in the air ensures that biological processes can be sustained for millions of years in the future.

Nitrogen

The presence of nitrogen and its compounds in the biosphere is another essential constituent for the maintenance of life. Exchanges of nitrogen take place within the biosphere constituting the nitrogen cycle, which has been described in Chapter 2. The nitrogen cycle is closely linked with the carbon and oxygen cycles, because all these cycles are part of the biological life processes involving oxidation, respiration, and the production of proteins. The cycles are interdependent in any one organism, and operate continuously throughout the biosphere.

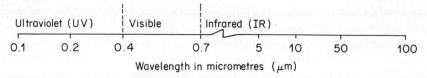

Figure 4.5. Regions of the electromagnetic spectrum.

4.3 Solar Radiation

Solar radiation from the sun passes through outer space, and through the atmosphere to the earth's surface. The radiation consists of energy in the form of electromagnetic waves of differing wave lengths (see Figure 4.5). Solar energy in the atmosphere consists of about 99% short waves (UV and visible light waves) and 1% long waves. Not all the energy is transmitted straight through the atmosphere, but some is absorbed and reflected in a complex pattern of energy movements, or the solar radiation cycle (Figure 4.6). Short wave radiation of below 0.4 μm length (UV) readily passes through the atmosphere, and about 20 units (per 100) reach the earth's surface direct. The other 80 units are absorbed by CO_2, O_2, and O_3 molecules, and water vapour in the atmosphere. Some of the absorbed energy (15) is retained and converted into heat, but the remainder (65) is either scattered or reflected by the atmospheric constituents. Scattered and reflected energy may either pass back into outer space (34), or to the earth's

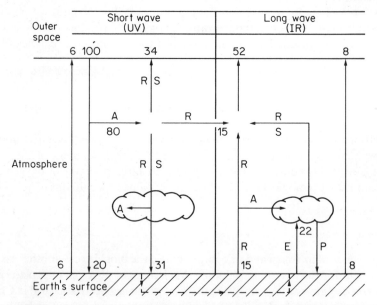

Figure 4.6. The annual solar radiation cycle: A, energy absorbed by atmospheric gases; R, energy reflected by atmosphere and clouds; S, energy scattered by dust and smoke particles; E, evaporation; P, precipitation. The numbers represent units of solar energy.

surface (31). A small amount of UV energy is reflected from the earth (6) directly back into space. The ozone layer in the atmosphere absorbs significant amounts of the short wave radiation, and so helps to protect the biosphere. Overall, if 100 units of UV radiation enters the atmosphere, only a net quantity of 45 units is received by the earth.

When short wave radiation is absorbed, heat energy is produced, and long wave IR radiation is evolved and emitted by radiation and convection. The earth's surface is an absorber and emitter of IR, and this provides the main heat source for the lower atmosphere. The emitted terrestrial IR energy has a wave length of 10–50 μm, and very little passes directly through the atmosphere into outer space (8). Most of the radiation is absorbed in the troposphere (37) particularly by water vapour, CO_2, O_3, and clouds. The lower atmosphere has therefore been likened to a 'heat blanket' which helps to provide a warm layer of air close to the earth. This absorbed radiation as heat is either reflected back into space (52 units), or to the earth, particularly from clouds, and is distributed by air movements. Evaporation of water from the large water surfaces on earth requires latent heat drawn from the earth. The water vapour produced rises into the air where it cools and condenses into clouds, releasing the latent heat into the atmosphere (22 units). In GB this contribution to the troposphere is high in summer, and about half the solar radiation reaching the earth is used for water evaporation. In addition to re-radiated solar energy as heat, the combustion process of burning fuels also emits heat into the lower atmosphere, particularly over large urban and industrial areas. Overall, if 45 units of UV solar energy are converted into heat on the earth, then about 60 units of IR are lost from the atmosphere, so consequently the atmospheric temperature is lower than the earth's over an annual period. It is also clear that any changes in the atmosphere that affect the way in which the solar energy cycle operates can cause changes in the earth's temperature and climate.

4.4 Atmospheric Temperature

Temperature variations occur throughout the atmosphere and there are different temperature gradients in the four atmospheric layers (Figure 4.7). From a pollution point of view, the temperature changes in the troposphere and lower stratosphere are significant. The normal temperature gradient in the troposphere is a decrease of 1°C for every 300 metres rise in height. In the stratosphere, the general trend is an increase in temperature with altitude. Scientific investigation using satellites is now being carried out in the stratosphere to increase general knowledge, and also to see the effects of jet-engined supersonic aircraft that now fly at altitudes of between 10 000 and 14 000 m. In the mesosphere, the temperature gradient is similar to that of the troposphere. Little is known about temperature changes in the thermosphere except that the temperature increases rapidly up to about 1000°C under the effect of solar radiation, and there is little absorption and scattering in the thin atmosphere.

The mean earth temperature only appears to vary about 1°C above and below

34

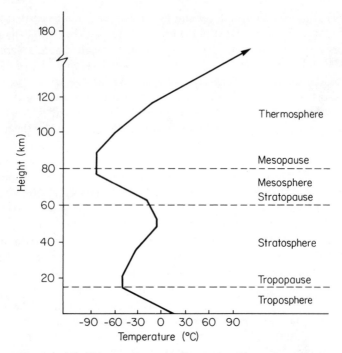

Figure 4.7. Temperature gradients in the atmosphere. Troposphere — decrease of temperature with height to 12 km. Stratosphere — gradual increase of temperature with height to 50 km. Mesosphere — decrease of temperature with height to 80 km. Thermosphere — increase of temperature with height, rapidly after about 110 km. (After diagram in *Introduction to Meteorology,* by S. Petterssen, McGraw-Hill, 1958.)

12°C. Earlier in the earth's history, between about 6000 and 3500 BC, the temperature probably rose some 2°C. The effect of this was considerable, as ice sheets melted and the polar ice caps retreated to the poles. These changes were accompanied by a significant rise in sea levels, and a redistribution of land masses by the formation of the Baltic and North Seas and the English Channel. Between 1900 and 1940 there was a rise of about 0.7°C in the average air temperature of the Northern Hemisphere (Figure 4.8), but between 1940 and 1960 the temperature dropped by approximately 0.6°C. It is thought that a temperature change of 1°C or more in the lower atmosphere could have significant climatic effects, so it is important to examine the causes of such changes.

It is known that CO_2 in the atmosphere does not absorb the incoming short wave UV radiation to any extent, but the gas strongly absorbs a band of long wave IR, particularly the earth's thermal radiation. This absorptive property of CO_2 prevents IR being emitted to the other atmospheric layers and lost into outer space. The warming effect of this upon the atmospheric temperature has been likened to a blanket or greenhouse effect. The 1900–1940 rise in air

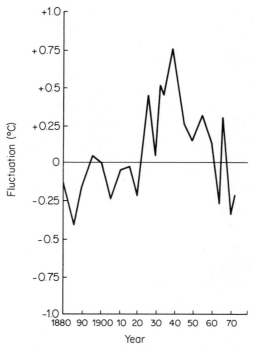

Figure 4.8. Air temperature fluctuations, 1880–1974.

temperature has been attributed to the greenhouse effect, because during the same period the atmospheric concentration of CO_2 was also increasing, and so more absorbed terrestrial radiation would be reflected back to the earth. But since 1940, whilst the CO_2 concentration has continued to rise, the average air temperature has fallen. The actual effect of the CO_2 increase in the lower atmosphere is therefore not clear, and is arguable. Some experts have estimated that a future increase of 20% in the CO_2 concentration to 384 ppm could raise the earth's mean temperature by 1°C, and cool the stratosphere by 0.5–1.0%. This could result in considerable ice melting, widespread flooding of low-lying coastal areas, and significant ecological and human effects. Whether this eventually will take place in the future it is impossible to predict at present. It is also important to realize that temperature changes in themselves not only affect the lower atmosphere, but also impinge upon all the natural cycles that exist in the biosphere.

4.5 Air Movements

Air is warmed by the absorption of terrestrially reflected solar radiation, as well as by heat from combustion processes on earth. A study in Sheffield showed that heat produced by combustion was about 20% of the total solar radiation received per year, or one third of the net radiation balance. The density of air is inversely proportional to temperature, and so as air is warmed its density is lowered, and it expands and rises. The rising air soon loses heat because of the troposphere temperature gradient, and so as the density increases it descends to the earth again. This upward and downward movement of air has importance in relation to pollution dispersal. The lower atmosphere also contains water vapour of varying amounts according to the hydrological cycle and the earth's temperature. Warm air occupies a larger volume than cooler air, and so is able to retain more water vapour. The air can become saturated, and as it cools the water vapour condenses to form clouds or mist or fog.

During conditions of daylight and solar radiation, warm air is usually rising and carrying up water vapour into the troposphere. During darkness, solar radiation is absent, and air near the earth's surface cools as heat is released to the ground. During two out of five nights in GB, a temperature inversion occurs at a height of about 3–350 metres above the earth, where the air is warmer than the ground layer of air beneath it. This has the effect of reducing upward air movements in the ground layer. If the temperature inversion is over urban and industrial areas with a high degree of humidity and pollution, then fog or smog conditions can occur. The resulting pollution and fog, or smog, tends to persist because it reduces solar radiation reaching the ground, the air temperature does not rise very much, and there is little air movement. Smog conditions occur especially in industrial valleys and river basin areas such as Los Angeles and Tokyo, which are well known for their frequent pollution problems. Inversions can occur at the higher altitude of 150–900 metres, and they can cause similar conditions where pollution is trapped, or is slow to disperse in the boundary layer of air close to the earth.

Another meteorological condition associated with industrial cities is the heat island. This occurs above tall buildings where there is a layer of warm air at about 5–7°C above ground temperature, sandwiched between two cooler layers of air above and below it. The heat island is often plateau-shaped, and is particularly distinctive in conditions of clear skies and little air movement at night. The temperature inversion condition reduces heat loss from the lower atmosphere and tends to retain pollution. The formation of heat islands is not fully understood, but may be associated with the thermal lag that occurs at night in large urban areas, when there is a slow release of absorbed heat from buildings and the earth's surface. Often heat islands disperse by mid-day and they are not considered to be a pollution hazard at present. However, the increasing consumption of energy may increase their importance and effect in the future.

There is a continuous emission of pollution in the form of gases, fine solids, and liquid droplets from chimneys, open fires, and motor vehicles. The pollution will disappear if it moves rapidly upwards and is dispersed and diluted in the troposphere. If this succeeds the hazardous constituents are reduced to harmless concentrations in the earth's boundary layer of air from 0 to 600 metres high, in which people live and breathe. Under optimum conditions, air pollution from a chimney is carried up by the velocity and temperature of the issuing discharge and turbulent rising air currents. At a height above the top of the chimney, the issuing plume diffuses and dilutes in the troposphere, due to horizontal winds of optimum speed (Figure 4.9 (a)). If a temperature inversion is present below the chimney height, then there are turbulent rising air conditions above the plume and the smoke and fumes are carried up and away from the ground (Figure 4.9 (b)). However, if the temperature inversion is above the chimney height, relatively stable air conditions exist which prevent upward movement of the plume. The unstable turbulent air below the inversion can cause downward dispersion and eddying of the emission in the boundary layer (Figure 4.9 (c)). This condition can also occur when there is no inversion present, and cloudy or heat island conditions again create unstable turbulent air near the ground. The air movement will then cause the polluted emission to eddy and swirl about, and again it could be blown towards the ground (Figure 4.9 (d)).

In these examples showing the effect of different climatic conditions, the emission from only one isolated chimney stack was considered. In the built environment there may be numerous chimneys and also neighbouring buildings of various heights. The effect of multiple buildings is to further complicate the

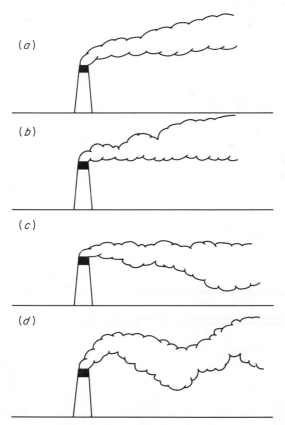

Figure 4.9. Effect of atmospheric conditions on chimney emissions.

atmospheric behaviour patterns, because buildings of different heights will obstruct and distort the air flow, and can produce calm or very turbulent mini-localized conditions. For example, there can be down-draughts on the lee side of chimneys or buildings, pockets of still, polluted air, and irregular dispersion of smoke and fumes. The irregular profile presented by buildings also produces a frictional effect upon the movement of the boundary air layer as a whole, affecting its velocity and direction of movement. Adverse ground level pollution conditions are a cause of public complaint. The complainants are often pacified by an agreement to increase chimney height. This modification can only be successful if the local climatic conditions are closely studied and known, especially in relation to the prevailing winds, patterns of air movements, and inversions. Note that altering the height of a chimney only changes the position of the emission of the pollution, and does not in itself reduce or remove it. To deal with pollution emissions satisfactorily, account must be taken of the quantity, density and height of the efflux, as well as the temperature and velocity of the air at the point of emission.

CHAPTER 5

The Extent of Atmospheric Pollution

5.1 The National Survey

Various estimates have been made of the total amount of pollution entering the atmosphere over the UK. The information is obtained by a process of monitoring that involves the sampling of pollutive constituents present in the air or deposited on the ground. Different types of monitoring are carried out by various agencies in the UK, and consequently the data available varies in respect of the type of pollutive constituents measured and the quantities recorded. The most reliable information showing the most complete coverage of air pollution is provided by the National Air Pollution Survey (NAPS). Under this,

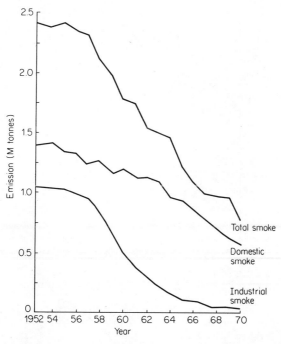

Figure 5.1. Smoke emission in the UK, 1952–1970 (from the *National Survey of Air Pollution Report,* 1970).

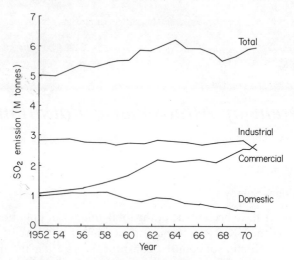

Figure 5.2. Sulphur dioxide emission in the UK, 1952–1971 (from the *National Survey of Air Pollution Report*, 1970).

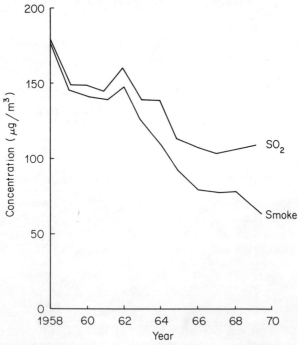

Figure 5.3. Average SO_2 and smoke concentration at ground level 1958–1970. (Adapted from the Royal Commission on Environmental Pollution *First Report*.)

Table 5.1. Annual output of pollution in the UK, 1970–71

Pollutant	Output (M tonnes)
Carbon monoxide	16.3
Oxides of sulphur	6.29
Carbon dioxide	6.2
Smoke	0.84
Grit and dust	0.55
Unburnt hydrocarbons	0.48
Aldehydes	0.26
Oxides of nitrogen	0.26
Lead	0.006
Total	31.186

regular daily sampling is carried out at 1200 sites throughout the country. All the sites measure smoke and sulphur dioxide concentrations, and the results are shown in Figures 5.1 to 5.3. Details of other types of monitoring for atmospheric pollution are included in Chapter 25. A more comprehensive estimate of the pollution output in the year 1970–71 is shown in Table 5.1.

The accuracy of the quantities given in the table could be argued, but the value of the estimate lies in the relative proportions of the different pollutants. In terms of quantity the most serious constituents are carbon monoxide, oxides of sulphur, carbon dioxide, and smoke. However, in respect of pollution effect on people, the most serious are oxides of sulphur and smoke, which substantiates the daily sampling of these by the National Survey. Carbon dioxide is not regarded as a pollution hazard at present, but the output into the global atmosphere is continually increasing, and is probably of the order of 500 000 M tonnes per year. Carbon monoxide does not generally reach toxic amounts in the UK at present, but its annual output is also increasing, largely because of the increasing numbers of motor vehicles in use.

5.2 The Main Sources of Pollution

The main sources of atmospheric pollution may be summarized as follows:

(a) The combustion of fuels to produce energy for heating and power. This is carried out in most industrial, commercial and domestic premises.
(b) The exhaust emissions from transport vehicles that use petrol, or diesel oil, or kerosine fuels, including road vehicles, diesel locomotives and aeroplanes.
(c) Waste gases, dust and heat from many industrial sites, including chemical manufacturers, iron and steel smelting works, cement and brick works, quarries, and electrical power generating stations.

A considerable amount of air pollution results from the burning or combustion of fuels. The comparative use of fuels is shown in Table 5.2. Fuels are primarily derived from fossilized plant material and consist mainly of carbon and or its compounds. They are burned in air and undergo combustion to produce heat

Table 5.2. Air pollution from fuels in the UK. (Adapted from *Clean Air Year Book* 1976)

Fuel	Type of user	Energy used (M tonnes)	Pollutants (M tonnes) Smoke	SO_2
Coal	Domestic	14.4	0.44	0.24
	Electrical power stations	79.9		2.04
	Industrial and miscellaneous	19.7	0.05	0.34
	Coke ovens	21.9		0.10
	Gas supply industry	0.5		0.01
Coke and smokeless fuels	Domestic	4.8		0.08
	Industrial and miscellaneous	3.0		0.05
Oil	Domestic	3.8		0.02
	Industrial and commercial	57.1		2.37
	Electrical power stations	17.0		0.95
	Petroleum refineries	3.9		0.22
Total as coal equivalent (tce)*		303.2		
Total			0.49	6.42
All fuels	Domestic	26.6 tce	0.44	0.34
	Industry—all users	165.9 tce	0.05	3.09
	Electrical power stations	110.6 tce		2.99

*1 tonne of oil = 1.7 tonnes coal equivalent.

energy, and gaseous and solid waste products. When combustion is complete the main gaseous product is CO_2, but if there is incomplete combustion and oxidation then carbon monoxide (CO) is produced, in addition to hydrocarbons. Incomplete combustion of coal produces smoke consisting of particles of soot or carbon, tarry droplets of unburnt hydrocarbons, and CO. Fossil fuels also contain 0.5–4.0% of sulphur, which is oxidized to SO_2 during combustion. Table 5.2 shows that all the smoke produced originated from coal. Domestic combustion appliances are obviously inefficient, because the sector used 8.8% of the total fuel, but produced 89.8% of the total smoke emission. Sulphur dioxide was produced from all types of fuel, to the extent of 2.0% by weight from coal, 1.7% from smokeless fuels, and 4.35% from oil. The domestic sector only produced 0.3 M tonnes of SO_2 (5.3%); so that industry, power stations, and transport were the major producers of this pollutant. It should also be noted that oxides of nitrogen are another pollutant produced by the combustion of all types of fuel, but no quantities are available for individual fuels.

The overall picture with regard to smoke and SO_2 is shown in Figures 5.1–5.3. Smoke emission has markedly decreased since about 1958, and this is no doubt due to the implementation of smoke control orders made under the Clean Air Acts, and changes in fuel utilization. Smokeless fuels have been used domesti-

Table 5.3. Energy Consumption in the UK by all users

Fuel	1960		1965		1976	
Coal	202M	tonnes	188M	tonnes	76M	tonnes
Town gas	2636M	therms	3338M	therms	47M	therms
Natural gas	—		—		6610	,,
Petroleum	38.5M	tonnes	60.5M	tonnes	107.5M	tonnes

cally instead of coal, and both the domestic and industrial sectors have switched from using coal to petroleum and natural gas as shown in Table 5.3. Between 1960 and 1976, the use of coal by all categories of users went down by 126 M tonnes per year or 62%, but the consumption of gas rose by 4021 M therms or 253%, and petroleum by 69 M tonnes or 279%. Consequently smoke emission in UK declined by 68%, from 2.4 M tonnes in 1955 to 0.775 M tonnes in 1970 (Figure 5.1). Some industries, such as power stations, cooking ovens for iron and steel making, and cement production, still use coal. However, the use of tall chimney stacks, stack gas cleaning using electrostatic precipitation, centrifugal collectors and gas washing has helped to reduce the emission of smoke and dust near to the ground (Figure 5.3). The position with regard to SO_2 pollution is much less satisfactory. The overall position is shown in Figure 5.2, and SO_2 emission in the UK increased slightly from 5 M tonnes in 1950 to nearly 6 M tonnes in 1970. Pollution from the domestic sector declined by 30% over the 20 year period, and this has reduced the SO_2 concentration at ground level. This fall is most probably due to the reduced consumption of coal, and smoke control orders. However, the SO_2 emission from industry remained about constant, but SO_2 output from power stations increased by 2 M tonnes per year between 1950 and 1970. Over this period, the proportion of primary fuels consumed for electricity generation increased from 18 to 27%, and electricity consumption doubled to 211.6 TWh (million million watt hours). Power stations clearly emit a significant amount of air pollution, and in 1970 they produced about half the total SO_2 emitted. They also produce waste heat and water vapour from the top of cooling towers and warm water from the bottom, which is often released into rivers causing heating up of the river water and eutrophication. Present industrial SO_2 pollution is increasing, and despite the fall in SO_2 concentration at ground level, the concentration higher in the troposphere is not. The present industrial policy for SO_2 is not removal, but dispersion into the air by discharge from tall chimney stacks. In fact a 1000 MW power station has to have a stack height of 244 metres to reduce the ground concentration of SO_2 to 0.14 ppm, and comply with the accepted limit. Therefore the tropospheric concentration of SO_2 is increasing in step with expanding industrial production and increasing demand for electrical energy. The effect of this pollutive SO_2 is not only confined to the UK, but it is international and global. It may well be affecting some of the natural biotic cycles. The total amount of sulphur in the natural cycle has been estimated at 142 M tonnes per year, but the contribution of man-made pollution is probably of the order of 73 M tonnes of sulphur per year.

Road Transport

Road vehicles are the second major source of air pollution, and a 1973 estimate is given in Table 5.4. The data given is based upon nearly 17 M vehicles. About 86% were cars and motor cycles using petrol, and 1.06 M or 11% were goods and passenger vehicles, which use mainly diesel fuel. Even allowing for the difference in the types of vehicle it is clear that diesel engines are much less polluting than petrol engines. Both types of engine are not very efficient convertors of fuel to energy. However, diesel types, with a conversion efficiency of around 30%, must be more efficient and use less fuel than petrol types with a 15–20% conversion efficiency. Both types of engine have incomplete combustion of fuel so the major pollutant is CO, amounting to 91% by weight of all vehicle emissions. Internal combustion engines also produce 0.05% unburnt hydrocarbons such as paraffins, olefins, polynuclear aromatics, and aldehydes. During combustion in the engine cylinder, reaction occurs between atmospheric oxygen and nitrogen to produce nitrogen monoxide (NO). This gas passes out of the engine in the exhaust phase, cools down, and then combines with more oxygen to form nitrogen dioxide (NO_2), and nitrogen tetroxide (N_2O_4). It is common practice to describe this mixture of compounds as oxides of nitrogen (NO_x). Small amounts of oxides of sulphur, mainly as SO_2, are produced by the oxidation of sulphur present in both petrol and diesel oil. The proportion of all these pollutants varies with the type of engine, its speed, maintenance, and operating efficiency. For example, the quantity of unburnt hydrocarbons emitted from a diesel engine is increased if its ignition is badly adjusted and maintained, and even some cars produce smoke.

The practice of using tetraethyl lead and other lead tetra-alkyls as additives in petrol was introduced in 1923. They were added to improve the octane rating of low quality petrol for use in high compression engines, and to prevent semi-explosive combustion or knocking in the engine. During combustion, the organic lead compounds decompose to produce inorganic lead, and about 60% of the lead fuel content passes into the air in the exhaust gases. In the UK in 1973, it was estimated that over 8100 tonnes of lead per year from petrol entered the atmosphere as a toxic pollutant. This figure was probably too low, as a 1977 estimate was 250 000 tonnes per year. Certain employees such as policemen,

Table 5.4.   Vehicle Pollution in the UK, 1973

Pollutant	Quantity (M tonnes)		
	Petrol-engined	Diesel-engined	Total
Carbon monoxide	8.0	0.12	8.12
Unburnt hydrocarbons	0.39	0.024	0.414
Oxides of nitrogen (NO_x)	0.25	0.08	0.33
Aldehydes	0.01	0.003	0.013
Oxides of sulphur (SO_x)	0.016	0.03	0.046
Lead	0.01	—	0.01
Total	8.666	0.257	8.923
	(97.1%)	(2.9%)	

garage workers, car park attendants, bus drivers, and highway workers are obviously at risk, and some workers have accumulated relatively high body lead concentrations. Similarly, residents living near motorways and interchanges are also at risk where there is a sustained high concentration of vehicle exhaust fumes. However, over most of the UK, the atmospheric lead content has not yet reached harmful levels. Only in localized situations is the lead concentration relatively high. In some multi-storey car parks in 1976, a lead concentration one hundred times above the average city street level was recorded at certain times of the day. The obvious way to lessen the risk of lead pollution is to stop using lead additives in petrol. They are not essential, and further refining of petroleum could produce satisfactory high octane fuel, but at a higher cost. In the early 1970s it was decided that some antipollution action should be taken for lead, because the risks would increase in the future as the number of vehicles in use continued to rise. The UK Government in 1974 reduced the level of lead permitted in petrol from 0.64 to 0.55 g/l. Further reductions to 0.50 g/l, as soon as possible, and to 0.45 g/l by 1978 have also been approved for the UK. The EEC have also been concerned about lead pollution, and their recommended levels of lead are 0.4, and ultimately 0.18 g/l. The UK has recently stated that it will reduce petrol lead content to 0.4 g/l in 1981, in harmony with the EEC standards.

The primary pollutants produced in vehicle emissions undergo a series of complex interrelated chemical reactions in the troposphere and lower stratosphere to form secondary products. These reactions utilize solar UV energy from the atmosphere, and so are called photochemical pollution reactions. The nature of the photochemical changes has been demonstrated in the laboratory, but there is some uncertainty regarding the precise reactions that occur in the atmosphere, and the quantities of secondary pollutants produced. Some of the reactions that are involved are briefly summarized in Table 5.5.

In areas of high traffic density and temperature inversion conditions such as in Los Angeles, it has been estimated that 762 tonnes of NO_x and 254 tonnes of organic substances per day are released into the atmosphere. The resulting photochemical smog consists of a mixture of primary and secondary pollutants, of which PAN and ozone concentrations are considered to be of significance. Both these substances appear to be hazardous, and it is known that exposure to 1 ppm of ozone for eight hours a day over a period of a year can cause diseases of the respiratory system. In Los Angeles it has been estimated that the ozone level may exceed 160 ppm on three days out of four. Photochemical smog has caused eye and throat irritation, reduced solar radiation, and caused plant foliage damage in both Los Angeles and Tokyo. In most towns in the UK there is no evidence, as yet, that photochemical smog is a pollution hazard under normal atmospheric conditions, and there is no evidence that it is affecting human health. However, in large urban areas such as the London Metropolitan area, photochemical smog is present in the summer over an area of about 48 square km. In July 1976, during exceptional heatwave conditions, the London ozone level rose to double the accepted threshold level of 0.1 ppm. PAN has also been detected in small amounts in London and Harwell.

Table 5.5. Photochemical pollution reactions

Primary pollutant	Reactions	Secondary pollutants
Nitrogen dioxide (NO_2)	(a) Dissociates to atomic oxygen and	Nitrogen monoxide (NO)
	(b) NO oxidized to.	Nitrogen dioxide
Molecular oxygen (O_2)	Dissociated to atomic oxygen (O)	
Hydrocarbons and aldehydes	(a) Dissociate to free organic radicals	
	(b) Radicals with O form peroxyacyl radicals	
	(c) Peroxyacyl radicals with NO_2 to	Peroxyl-acetyl-nitrate (PAN)
	(d) Peroxyacyl radicals with O_2 to	Ozone (O_3)
	(e) Free radicals with oxidants to	Aldehydes, ketones and peroxybenzol
Carbon monoxide	Oxidized to	Carbon dioxide

In the short term, the concentration of smog constituents in the UK needs careful monitoring, as the amounts of some substances may be approaching threshold limit values in large towns, in some months of the year. In the long term, the health effect of photochemical constituents upon the population needs medical investigation. Measures should be taken to reduce the output of primary pollutants from road vehicles. Theoretically, changes in engine design by, say, improved injection and combustion of fuel in the cylinders, or combustion of unburnt gases in the exhaust manifold, or catalytic conversion in the exhaust system, could improve the engine efficiency, and decrease the exhaust emission of hazardous pollutants. Car manufacturing regulations in the US are now in force to reduce the exhaust pollutant emissions in new models by 90%, and UK manufacturers who export to the US are complying. It is considered unlikely that there will be a large decrease in the number of road vehicles in use in the UK over the next two decades. If more of these vehicles were powered by diesel engines in preference to petrol, and if others did not use a petroleum fuel at all but used methane, or hydrogen, or electric motive power, then there would probably be no future danger from photochemical smog.

Industry

The third main source of air pollution is from industrial premises and electrical generating power stations. All works and power stations that use the combustion of coal, coke, or petroleum for the generation of heat and power produce smoke and SO_2. In 1969–70 they produced 50.5% by weight of the total output of smoke and SO_2 in UK (Table 5.2). However, industry generally does attempt to reduce smoke and dust pollution, and between 1956 and 1973 the industrial smoke emission decreased by 96%, despite a considerable increase in the use of primary energy fuels, particularly petroleum. Smoke pollution from the domestic sector has been largely reduced by the use of smokeless fuels or the non-use of solid fuels, and industry employs smoke particulate and dust removal techniques. The combustion gas stream or flue gases can be treated by three possible methods before emission from the chimney stack. In the electrostatic precipitation method, the gases pass through an intense electric field where the particulates become charged and attracted to collector electrodes. The accumulated solids are then mechanically dislodged and collected. A new electrostatic precipitation installation can be up to 99.7% efficient. Some industries use cyclones or centrifugal collectors where the flue gases are forced up a helical path in a conical-shaped collector. The solid particles are spun to the cyclone wall by centrifugal force, and collected in a dust hopper at the base of the cone. The third method of gas scrubbing is widely used, and jets of water are directed into the gas flow and wash out the particulates.

The reduction of oxides of sulphur emissions by industry and power stations is still a problem. Theoretically there are two broad approaches, which involve either removal of the SO_x from gas discharges after combustion but before emission into the air, or reduction or removal of sulphur from the primary fuel before or during combustion. At present there is no satisfactory and practical way for removing SO_x and fluorine compounds from flue gases. One method of desulphurization is by wet scrubbing and limestone absorption treatment, but the process lowers the temperature of the gas stream. Consequently, upon emission from the stack it does not rise readily and disperse in the atmosphere. The process is in operation at two London power stations, where the annual capital cost is put at £ 85–170, and the operational costs at £ 25 per tonne of sulphur removed. The Central Electricity Generating Board (CEGB) has carried out desulphurization research for its power station pollution abatement programme. Techniques investigated include physical treatment of flue gases by wet scrubbing and active carbon adsorption, as well as chemical treatment. The latter involves oxidation of SO_2 to sulphur trioxide and subsequent reduction to sulphur. This process has the advantage of not cooling the flue gases, and the sulphur obtained can be sold as a useful recycled product to offset the costs of its removal. The alternative approach to desulphurization involves the treatment of coal or petroleum before or during combustion. Removal of sulphur from petroleum during refining is possible, but the cost has so far deterred development. A 1973 estimate for sulphur removal from fuel used in Western Europe was £ 1200 M for capital investment, and the operational costs would require a 25% increase in fuel prices. The National Coal Board (NCB) in conjunction with the EEC is developing a fluid bed combustion technique for steam raising. This process includes sulphur removal, it can use coal of widely varying quality more efficiently, and produces a reduced amount of NO_x. Once in industrial use, fluid bed combustion could significantly reduce the air pollution produced by large industrial plants and power stations, perhaps in the 1980s. Until the sulphur content of fuels is reduced or removed by some of these possible treatment processes, then the SO_2 pollution of the atmosphere will continue.

In addition to smoke and SO_2 many industrial manufacturing processes are potential sources of numerous other air pollutants, including toxic gases, heavy metals, complex organic compounds which may have carcinogenic effects, as well as dust, and grit. The industrial plants that appear to be the major sources of pollution include iron and steel foundries, coking ovens, pulp and paper mills, petroleum refineries, chemical plants, and cement works. Other plants may cause localized pollution problems but they are not sources of general pollution importance. There are no readily available national statistics for total UK industrial pollution. Under the Alkali Act 1906, registered processes must control pollution and supply emission data to HM Alkali and Clean Air Inspectorate, but these data are not available to the public. The Control of Pollution Act 1974 requires non-registered works to supply emission data to Local Authorities who may be required to publish it, but the Act is not yet fully operative. Consequently the public have no overall pollution data available to enable them to assess the extent of air pollution, or the effectiveness of pollution statutes and their implementation.

There has been a considerable improvement in the emission of dust into the atmosphere, as more industrial processes have come within the scope of the Alkali Act. Examples of the changes in just three industries, which were scheduled in 1958, are shown in Table 5.6. These examples serve to show the more effective control of pollution by HM Alkali and Clean Air Inspectorate who implement a national policy of agreed standards. There is little doubt that the Alkali Inspectorate system for registered processes generally works efficiently, and many potentially toxic pollutants are prevented from entering the atmosphere. On the other hand the effectiveness of control over pollution from non-registered works is not known but may be suspect, because Local Authorities vary in their attitudes and effectiveness. However, these Authorities work within their local areas, and so are much more open to public pressure and complaints concerning local incidence of air pollution. The DOE Central Unit for Environmental Pollution has proposed that a more systematic monitoring of airborne metal discharges needs to be undertaken. At present only local and occasional regional surveys have been made for specific metals. The Unit

Table 5.6. Emission of dust into the air, 1958 and 1972

Industry		Production (M tonnes)	Increase (%)	Dust emission to air (tonnes)	Reduction (%)
Cement	1958	10		160 000	
	1972	16	60	37 000	64
Sinter plants	1958	8.5		130 000	
	1972	20	135	40 000	69
Power stations*	1958	43		1.0M	
	1972	59	35	189 000	82

*These figures refer to coal burnt.

propose a network of twenty monitoring sites to carry out monthly sampling and assaying of lead, arsenic, cadmium, copper, mercury, zinc, vanadium, germanium, and selenium. This proposal should be carried into operation as soon as possible. It is only through regular and standardized monitoring of atmospheric pollution that its extent and hazardous concentrations can be observed.

5.3 Radioactivity

All chemical substances are composed of atoms, which contain a central core or nucleus surrounded by electron particles speeding round the nucleus in circular or elliptical orbits. The nucleus is composed of proton and neutron particles which constitute most of the mass of the atom. The ninety-two different types of naturally occurring atoms and elements differ from one another in respect of the number of protons and neutrons in the nucleus, and the number of orbiting electrons. Atoms and elements with identical chemical properties, and with nuclei containing the same number of protons but differing numbers of neutrons, are called isotopes or varieties of the same chemical element. For example, oxygen occurs naturally as three isotopes, expressed chemically as 0–16, 0–17, and 0–18, where the letter indicates the kind of element and the figures indicate the mass number (protons plus neutrons) of the atom.

If an atomic nucleus is unstable it may undergo spontaneous disintegration or decay. This process produces radiation, and the atoms that undergo disintegration are said to be radioactive. Radioactivity is shown by some 18 types of naturally occurring elements, but all elements are known to become radioactive if processed in a nuclear reactor, cyclotron, or other high energy neutron device. Radioactive atoms produce radiation continually, and this can be measured in curies, where one curie is the radioactivity of one gram of radium or 3.7×10^{10} disintegrations per second. For any one type of radioactive atom, the decay process occurs at a constant rate over a specific period of time. The decay rate is expressed in terms of the 'half-life', which is defined as the period of time within which half the nuclei in a sample of radioactive material undergo decay. For example, if a sample of strontium-90 contains 1000 atomic nuclei, then 500 will decay over a period of 28 years, and a further 250 of the 500 nuclei remaining will decay over the next 28 years, and so on. So the half-life of strontium-90 is 28 years. Half-lives of different radioisotopes range from parts of a millionth of a second to millions of years. This decay process and the rate of emission of radioactivity cannot be altered or prevented. Once a radioactive substance is released into the environment, it continues to give out radiations for a period of time commensurate with its half-life, until all the atomic nuclei have disintegrated. The types of nuclides, or nuclear species that may be released into the environment are shown in Table 5.7. The nuclear disintegration produces so-called radiation that consists of alpha and beta particles, gamma rays, and neutrons. Alpha particles have a positive charge and consist of two protons and two neutrons, and beta particles are high energy electrons with a negative charge. Gamma rays are high energy electromagnetic radiations that behave in a similar

Table 5.7. Some environmentally important nuclides

Source	Nuclide symbol	Nuclide name	Type of radiation	Half-life
Background radiation	Rn	Radon-222	alpha	3.8 day
	Ra	Radium-226	,,	1622 yrs
	C	Carbon-14	beta	5570 yrs
	K	Potassium-40	,,	1.3×10^9 yrs
	U	Uranium-238	alpha	4.5×10^9 yrs
	Th	Thorium-232	,,	1.4×10^{10} yrs
Nuclear weapon testing	I	Iodine-131	beta/gamma	8.1 days
	Sr	Strontium-89	beta	53 days
	Sr	Strontium-90	,,	28 yrs
	Cs	Cesium-137	beta/gamma	30 yrs
	C	Carbon-14	beta	5570 yrs
Nuclear reactors and airborne discharges	Ar	Argon-41	beta/gamma	1.8 hours
	Xe	Xenon-133	,,	5.3 days
	I	Iodine-131	,,	8.1 days
	Kr	Krypton-85	,,	10.8 yrs
	H	Tritium-3	,,	12.3 yrs
	C	Carbon-14	beta	5570 yrs
Fuel reprocessing and airborne discharges	I	Iodine-131	,,	8.1 days
	Kr	Krypton-85	,,	10.8 yrs
	H	Tritium-3	,,	12.3 yrs
Others	Cm	Curium-244	alpha/neutron	17.6 yrs
	Pu	Plutonium-238	alpha	86.4 yrs
	Am	Americium-241	,,	458 yrs
	Pu	Plutonium-240	,,	6.6×10^3 yrs
	Am	Americium-243	,,	78×10^3 yrs
	Pu	Plutonium-239	,,	2.4×10^4 yrs
	Pu	Plutonium-242	,,	3.8×10^5 yrs
	Np	Neptunium-237	,,	2.2×10^6 yrs

manner to X-rays. Gamma rays, like X-rays, are not stopped by soft tissues, and so constitute a considerable biological hazard. Neutrons are uncharged particles released from atomic nuclei. They are slowed down by collision with other atoms and molecules, and are largely 'captured' by their nuclei and produce further disintegrations as a result. Free neutrons do not exist as a long term radiation hazard, and are only present in close proximity to a nuclear reaction or explosion.

The first unit used to measure radioactivity was the curie (Ci), but this unit is too large for biological purposes and so two other units are commonly used. The physical unit used to measure the 'radiation absorbed dose' is the rad, and this is defined as 0.01 joules per kilogram, which is the amount of energy absorbed in a unit mass of material. A second unit is used to measure the radiation dosage received and its effect. This unit is derived from the X-ray dosage or roentgen (r), and it is called the 'roentgen equivalent man' or rem. The rem provides a biological measurement of the product of the radiation dosage in rads, plus a quality factor to take account of the effects of the different types of radiation. For example, a dosage of 5 rem could consist of 5 rad gamma radiation, or 0.5 rad of neutrons. Therefore the prescribed human dose limit is measured in rem, because this unit takes account of the biological effects of the different types of radiation. In this book all radiation doses are expressed in rem or millirem (mrem is one thousandth part of a rem) for convenience and simplification.

Atmospheric radiation emanates from both natural and man-made sources. The entire global population receives so-called background radiation from three natural sources in the environment (see Table 5.8). Cosmic rays from outer space enter the atmosphere, and the amount received varies with altitude and latitude. For example a population living at 1.5 km altitude in the higher latitudes will receive nearly twice as much cosmic radiation as a community living at sea level on the equator. The earth's crust also contains nuclides which continually emit radiation. Some examples are uranium-238, thorium-234, radium-223, and radium-226, which are present in rocks, soil, and natural building materials. Food grown in the earth, and drinking water which has percolated through soil, also contain nuclides. These are ingested into the human body in food and drink, and examples are potassium-40, carbon-14, and radon-222. This level of background radiation is not necessarily harmless, but it appears to be tolerated by the body and does not cause widespread tissue and genetic damage. There is some evidence that an above average background radiation can cause slight harmful effects. However there is no real evidence that the average natural background radiation of about 105 mrem per year is harmful in the long term. Genetic mutations are known to occur in the human population, but the causative agents may be chemical, or radiation, or both.

There are three types of man-made radiation sources. These are X-rays used in medicine and dentistry, fall-out resulting from nuclear weapons testing, and industrial emissions mainly from nuclear reactors and processing installations. X-rays are as penetrative to the body as radiation gamma rays, and they provide the greatest man-made radiation contribution. The average dosage per head of

Table 5.8. Average Background of natural radiation in the UK, 1976

Radiation source	Dose per year (mrem)	
	Bone marrow	Reproductive cells
Cosmic rays	33	33
Rocks, soil and airborne	44	44
Within the body	24	28
Total	101	105

N. B. The background radiation dosage can vary from 50 to 200 mrem per year according to altitude above sea level, latitude and geological rock formations.

population for X-rays was estimated as 32 mrem per year in 1976. When used for diagnostic purposes and radiotherapy they are obviously beneficial to the individual, but they should be used as sparingly as possible with adequate shielding of the reproductive organs or gonads. A United Nations Committee stated in 1972 that radiation from medical sources in each generation accounts for 120 cases of disease resulting from genetic damage. The Committee also added that the genetically significant dose from medical irradiation 'should not be allowed to increase unduly', and 'some kind of continuing appraisal of its magnitude is necessary'. The effect of X-ray exposure is cumulative in the body, and consequently the total dosage received by any individual should be carefully recorded and monitored.

The next largest man-made radiation contribution that the population receives probably comes from the atmospheric fall-out, produced by the explosive tests of nuclear weapons. Tests have been conducted in the air, on the ground, underground, and under the sea; and all, with the exception of underground tests, produced significant quantities of nuclear fission products such as carbon-14, strontium-90, iodine-131, and caesium-137. The radioactive material is dispersed in the atmosphere according to the size of the weapon. Nuclides from high-yield megatonne weapons enter the stratosphere, where they can remain for up to 10 years, but low-yield kilotonne detonations deposit nuclides into the troposphere for a period of only weeks or months. The nuclides may undergo further disintegrations in the atmosphere where they are globally dispersed, but eventually they are deposited upon the earth's surface as radioactive fall-out. The period of nuclear weapon testing can be conveniently divided into three phases. The post-Second World War above-ground tests commenced in 1945 and continued at periodic intervals until 1952. During this 7 year period, the US, USSR, and the UK used uranium-235 and plutonium-239 as the fission material in a total of 38 atmospheric tests. The second phase of testing was from 1952 to 1962, during which mainly thermonuclear or H-weapons were used for 393 detonations. Public concern about the build-up of radioactive contamination in the atmosphere grew during the late 1950s and early 1960s. Partly as a result of this pressure, the three nuclear powers signed a Partial Nuclear Test Ban Treaty in 1963, whereby they agreed to use only underground tests in the future. Since 1963, the US, USSR, and the UK have only carried out underground tests which produce little or no fall-out. France, China and India commenced testing nuclear weapons in the atmosphere in 1960, 1964, and 1967 respectively. They were not signatories to the 1963 Treaty, and detonated 48 atmospheric explosions between 1963 and 1973. It is difficult to find precise estimates of the amount of man-made radiation produced into the atmosphere from nuclear weapon tests. The peak year was 1962, when probably about 12 mrem was produced. There is also the long term situation to consider, and the period when the stratospheric radiation is coming down as ground fall-out. One estimate of this position is that the average dose of radiation that will accrue between 1968 and 1998 is about 140 mrem, caused by tests conducted before 1968. It is too soon to assess the long term effects of this radiation, whether they are somatic, or genetic affecting future generations. Perhaps the best that can be stated is that this source of atmospheric pollution has diminished since 1963, but it could easily increase again unless all nations that develop the nuclear weapon capability are signatories to a total test ban treaty in the future.

The third man-made source of radiation is from the increasing use of radio-isotopes in industry, research, and medicine, and from the use of nuclear reactors for power and electricity generation. The quantity of nuclides used for other than nuclear reactors is relatively small, and so their contribution to environmental radiation is not significant. The largest contribution arises from the operation of nuclear reactors, and the complementary nuclear fuel processing. The production of nuclear power may be divided into three types of operations. British reactors use uranium-238 ore concentrate as the basic raw material. This has to be preprocessed by purification, chemical conversions, and enrichment with uranium-235, and then manufactured into fuel elements. These consist of the nuclear fission material enclosed in an outer metallic cladding. The fabrication of the elements produces solid, liquid, and gaseous wastes, but they have a low total radioactivity. The second operation occurs in the reactor itself, where fission of the nuclear fuel releases heat energy which is used to drive a steam turbine electrical generator. Radioactive emissions to the air vary with the type of reactor. They include nuclear fission products, neutron generated nuclides induced in the coolant medium, and the products of corrosion and erosion processes in the reactor circuits. The gaseous wastes consist of seven nuclides which are detailed in Table 5.7. Liquid wastes containing tritium-3 and radioactive isotopes of iron, cobalt, and other corrosion products are also produced. Periodically the reactor has to be recharged with new fuel and the spent fuel elements are removed. These are extremely radioactive and must be stored under water for several months. In the third operation the spent fuel elements are reprocessed. This involves chemical treatment to separate the re-usable components of uranium-238 and -235, and plutonium-239 from waste fission products, cladding, etc. Gaseous waste products such as tritium-3, iodine-131, and krypton-85 are produced, as well as large quantities of liquid waste containing long half-life isotopes and some high activity solid wastes.

From the above summary it is clear that various types of nuclear reactor wastes are produced containing nuclides of varying quantities and radioactivity levels. Some gaseous nuclides are produced, and they are subject to stack filtration and absorption treatment before being released into the environment. It appears that not all nuclides are removed from the flue emissions, and new methods of treatment for iodine-129, krypton-85, tritium-3, and carbon-14 are being investigated. The estimated output of man-made radiation is shown in Table 5.9. These figures should only be taken as a rough estimate because accurate data are not readily available. It is probable that the global radiation level is of the order of 16 to 43 mrem per year. Man-made and background radiation together may be about 155 mrem per year, so that the former is only 24% of the total radiation to which the population is exposed. Also the contribution from nuclear tests and nuclear installations is only about 4% of the

Table 5.9. Estimated average man-made radiation in the UK, 1976

Radiation source	Dose per year (mrem)	
	Bone marrow	Reproductive cells
Medical—mainly X-rays	46	19
Nuclear test fall-out	6	4
Discharges from nuclear industry	0.25	0.2
Occupational	0.7	0.6
Total	52.95	23.8

total. The above is an estimate of the present position, but increased future use of nuclear fission reactors and nuclear energy will increase the environmental man-made radiation pollution. Radiation poses a new dimension to the environmental pollution threat of the future. Non-radiation pollutants and low half-life nuclides are being released into the environment, and under the present policy it is assumed that they will disperse and degrade in a relatively short period of time. Consequently they are not regarded as a future pollution threat until their volume and concentration greatly increases. But long half-life nuclides in nuclear power wastes will remain dangerously radioactive for hundreds and thousands of years. The process of nuclear decay cannot be accelerated, and these nuclides must therefore be retained within the environment for generations to come. The safe storage of these nuclides is as yet an unsolved long term problem, and the greater the quantity the greater the disposal problem.

An additional hazard is the possibility of accidental leakage of radiation into the environment from an accident in a nuclear establishment. Some isolated incidents have taken place in the UK, US, and USSR, but fortunately the pollution produced has usually been localized and of short duration. In October 1957, there was a reactor fire at Windscale, Cumbria, caused by a malfunctioning of the reactor temperature control system. Radioactive gas was released for a short period, and this moved downwind causing fission products to be deposited on the ground. The effect of this was the contamination of grass, which affected grazing animals and foodstuffs. The chief pollutant of the Windscale accident was iodine-131 (half-life 8 days), and to protect the population from contamination it was necessary to ban the use of all milk produced by cows in a 500 km^2 area, for several weeks. One of the most recent accidents took place at the Three Mile Island nuclear site at Harrisburg, Pennsylvania, in March 1979. This was caused by technical failures in the pressurized water system, so that the reactor overheated and released radioactive cooling water into the Susquehanna river, and contamination into the atmosphere. In both these accidents the extent of the pollution was low, and so no immediate deaths were caused. Both reactors were shut down, but the extremely high radioactivity resulting from the partial melting of the core fuel presented a future hazard. At Windscale, the reactor did not become operational again, but there remains the possibility of radioactive leakage. At Harrisburg, the authorities are hoping to operate the reactor again in

4 or 5 years' time, if it can be safely decontaminated and modified. A much more serious accident is thought to have occurred in the USSR in 1957. No official information has ever been released, but it is almost certain that a major nuclear waste accident took place near Kyshtym in the Urals. An area of about 3900 km^2 was probably so severely contaminated that it will remain uninhabitable for a very long time. Unlike the two other accidents, it is highly probable that many local inhabitants suffered radiation sickness, and there were possibly numerous deaths. It is the fear of an accident of this type that is causing anti-nuclear protests and demonstrations in a number of European countries and the United States. Whether the possibility of nuclear plant accidents is more probable than from some chemical plants, such as at Flixborough in 1974, is a matter of debate. However it is certain that nuclear reactor contamination can be more widespread, and have very long lasting effects upon the environment.

CHAPTER 6

The Effects of Atmospheric Pollution

6.1 Particulate effects

The pollution constituents that are emitted into the atmosphere are either gases or particulates. The latter consist of fine solids, or liquid droplets suspended in air. The larger sized particulates are grit, fly ash, dust, and soot, and the smaller sizes are smoke, mist, and aerosols. These different types of particulates have definitive meanings, as follows:

(a) grit—solid particles suspended in air with a diameter over 500 μm;

(b) dust—solid particles suspended in air with a diameter between 0.25 and 500 μm;

(c) smoke—gas-borne solids with particles usually less than 2 μm in diameter;

(d) fumes—suspended solids in air less than 1.0 μm in diameter, normally released from chemical or metallurgical processes;

(e) mist—liquid droplets suspended in air with a diameter of less than 2.0 μm;

(f) aerosol—solid or liquid particles in suspension in air, or some other gas, with a diameter of less than 1.0 μm.

Specific particulate size is related to the behaviour of pollutants and the time they reside in the atmosphere. Large solid particles with a diameter of over 50 μm are collectively visible in the air and settle out fairly quickly, so that they are not a long term pollution hazard. However, they often produce ground pollution because the larger sizes over 10 μm diameter are deposited near their point of emission. Consequently, stone and clay quarries, cement works, and brick works often cause despoilation of the surrounding land. Particulates in the size range of 50–0.01 μm diameter are of most significance in air pollution, and they are not obviously visible. They can remain in the atmosphere for varying lengths of time, and undergo chemical reactions to produce secondary pollutants. Particles below a diameter of 10 μm are able to act as nuclei for the formation of condensation water droplets in cloud formation. They may then be washed out of the air by precipitation as rain within about 7 days of their emission. The smaller solid and gaseous particulates can remain suspended in the atmosphere for days, weeks, months, or years. For example, in the lower troposhere for 6 to 14 days; in the upper troposhere for 2 to 4 weeks; in the lower stratosphere for up to 6 months; in the upper stratosphere for 1 to 3 years; and in the mesosphere for up to 5 to 10 years.

Atmospheric pollution is present largely in the troposphere and lower stratosphere, with a maximum concentration of small particulates of diameter 0.1 to 1 μm at about 18 km altitude. The atmosphere can be arbitrarily divided into four layers according to the behaviour and concentration of the pollution. Near the ground is a layer of air 1–100 m high which may be very polluted in local urban areas. Some of the pollution is absorbed into vegetation, buildings, and water surfaces, for example it has been estimated that 33% of the sulphur dioxide (SO_2) in this layer is removed by absorption. The next layer extends from 100 m up to the cloud base, at a height of between 500 and 2000 m altitude in the troposphere. Here the pollutants become well mixed by the turbulent air currents, and further amounts are washed out by drizzle, rain, and fog. The third layer contains most of the atmospheric water vapour and clouds, and extends up to the tropopause. Some pollutants may be dissolved or become nuclei in the cloud water droplets. Later they may either be removed from the layer as rain, or released again into the atmosphere when clouds evaporate. Small sized particulates remain suspended in the air and undergo photochemical changes using UV energy. There is also movement of particles by air currents, upward into the lower stratosphere, and horizontally over varying distances, depending on climatic conditions. There are well-known examples of dust and SO_2 transference between Europe and Africa and Scandinavia. The last atmospheric layer with significant pollution present is the lower stratosphere. This contains hardly any clouds or water vapour and a low concentration of pollutants. It is known to contain dust by direct upward emission from volcanic eruptions, ground and air nuclear explosions; and the products of photochemical reactions involving methane and sulphur trioxide. The general behaviour of pollution in the troposphere is a pattern of dynamic movement of the constituents, chemical changes, dispersion and dilution, and fall-out on to the earth. However, in the stratosphere, photochemical reactions occur but there is very little movement of the pollutants, and they remain in the layer for a very long period.

The high concentration of particulates in the atmosphere over large urban and industrial areas can produce a number of general effects. Smoke and fumes can increase the atmospheric turbidity and reduce the amount of solar radiation reaching the ground. Measurements over 60 years in some US cities have shown a 57% increase in turbidity, and in Switzerland there has been an increase of 88% over 30 years. Particulates can absorb and reflect incoming solar radiation and cause a 15–20% decrease in large cities compared to 'cleaner' rural areas. Readings of bright sunshine duration in central London between 1921 and 1950 showed that in winter the values were 50% lower, and in summer 10% below, those recorded in the surrounding countryside. Solid particulates take part in cloud formation, and so urban pollution and increased water vapour emission can produce up to 10% increased cloud cover, up to 10% more wet days, and increased mist, fog, and smog compared to non-industrial areas. These processes all combine to increase the deposits of large sized particulates on the ground. This in time affects the erosion and corrosion of buildings, materials, and metals, and also plant life. The part played by particulates in the air in relation to IR radiation and temperature changes has already been described. All these processes only occur in localized areas of heavy pollution. If the degree of pollution is less, or if it can be reduced, then air movements in the atmosphere will usually disperse and absorb the pollution satisfactorily.

6.2 Effects in the Biosphere

Chemical substances discharged into the atmosphere are often called primary pollutants. Some of these undergo chemical changes in the presence of water, oxygen, and ultraviolet solar energy to form intermediate or secondary products. Pollution effects in the biosphere may be caused by primary or secondary products. Some of the chemical changes that occur are briefly summarized in Table 6.1. The overall effect of air pollution upon the biosphere and the built environment can be broadly considered under three headings. The effect upon buildings and materials; the effect upon the soil, vegetation, crops, and animal life; and the effect upon people.

The fabric of buildings, that are surrounded by heavily polluted air for years, undergoes chemical changes. Gradual erosion takes place, and this is only too evident when the grimy upper surface is removed. There are many examples of the erosion of carvings which have become quite unrecognizable over a period of hundreds of years. The process is more rapid on the softer stone surfaces, where there is the combined effect of chemical pollution, wind and rain, and suspended dust and grit particles. Gaseous pollutants in the air, such as SO_2, CO_2, hydrogen sulphide, and fluorides, together with rain and solar radiation, attack the surface

Table 6.1.   Acid-forming reactions in the troposphere

Primary pollutant	Reactions	Secondary pollutants
Sulphur dioxide	(a) With water forms sulphurous acid	
	(b) Sulphurous acid oxidized to......	Sulphuric acid
Sulphur dioxide	(a) Slowly oxidized to sulphur trioxide	
	(b) Sulphur trioxide with water to	Sulphuric acid
Nitrogen dioxide	(a) With water to form	Nitrous and nitric acids
	(b) Nitrous acid with water to.......	Nitric acid and nitrogen monoxide
Carbon dioxide	With water to.................	Carbonic acid
Hydrogen sulphide	Oxidized to...................	Sulphur dioxide
Hydrogen fluoride	With water to................	Hydrofluoric acid
Silicon tetra-fluoride	With water to................	Hydrofluoric acid and SO_2

of buildings, particularly those composed of limestone and sandstone. Sulphur dioxide is converted into dilute sulphuric acid, which acts on calcium and magnesium carbonates (limestones) to form sulphates. These form a hard surface skin which blisters and scales off. In sandstone, the calcite cement is attacked by sulphuric acid, and the silica sand grains are loosened and washed away by the combined action of wind and rain. Stone erosion is also increased by CO_2, which dissolves in rain to form weak carbonic acid, and this attacks the stone to form soluble calcium and magnesium hydrogen carbonates.

Metal surfaces are attacked by atmospheric pollutants and undergo corrosion. The surface of iron and steel, in the presence of moisture and oxygen, changes into hydrated iron (III) oxide or rust, and the process is aided by SO_2 and CO_2. Copper and its alloys becomes coated with verdigris, or basic copper (II) sulphate in the presence of moisture, oxygen, and SO_2. Natural oxidation and corrosion of metallic surfaces occurs in unpolluted air, but the effect is greatly increased by pollution. Natural materials such as cotton, wool, and linen made up as dyed fabrics are also affected by pollutants. Sulphur dioxide dissolved in droplets of water forms sulphurous acid which can bleach or reduce the chemical dystuffs. The action is accelerated by ozone, and if dilute sulphuric acid is also present the fabrics will eventually rot and disintegrate. Plastic materials are more resistant to pollution, but some polymers eventually crack and become brittle. Leather and rubber also absorb sulphurous acid and are affected by ozone, so that they deteriorate over a period of time. Hydrogen sulphide in town air affects some paints which contain basic lead (II) carbonate pigment, so that they become discoloured when lead (II) sulphide is formed. Other pollutants also affect paint and cause the surface to break up and flake off.

The presence of gaseous pollutants in the air and the deposition of particulants on to soil can affect plants. The precipitation of rain, containing dissolved SO_2, over a period of time tends to lower the soil pH and it becomes acid. In addition the emission of heavy metallic compounds in localized industrial areas can cause toxic concentrations in the soil. Where these occur the range of plant species that can grow is severely restricted. Some species are very sensitive to small toxic amounts which inhibit the action of some plant enzymes. Particulates such as soot and dust are deposited on the leaves and block the stomata or pores of plants. This restricts the absorption of CO_2 and so reduces the rate of photosynthesis, as well as restricting the loss of water vapour and the transpiration rate. Both physiological effects cause stunting of the growth of all plants and reduce the yield of crop plants. Observations in Leeds show reduced growth of lettuce and radishes in heavily polluted areas compared to the less polluted areas of the city. In California, in conditions of very heavy pollution, leaf damage has occurred caused by the concentrations of SO_2, oxides of nitrogen, and ozone. In the Sacramento Valley in California, the combination of SO_2 and metallic pollutants killed all vegetation in an area of 260 km^2, and affected growth on a further 320 km^2 of land. Reduced yields of up to 30% have occurred and the cost has been assessed at $100 M per year. Alternatively the increasing concentration of CO_2 in the atmosphere might be expected to increase the rate of photosynthesis in plants, and provide a pollution bonus. This could be happening in unpolluted rural areas, but this effect will be more than offset by the other harmful results in heavily polluted areas.

Farm cattle can also be affected by air pollution. An isolated example occurred when smog conditions coincided with the 1957 Smithfield Fatstock Show in London. Beef and dairy cattle developed breathing difficulties and many died, but there was little effect upon sheep and pigs. Cattle grazing near aluminium and brick works have been known to show fluorine effects from the herbage, when they lose appetite and show a low milk yield, lameness, and joint stiffness. The salts of heavy metals from industrial plants can also cause toxic symptoms in cattle.

The general and most widespread effects of air pollution on people are caused by smoke and SO_2. These two pollutants are at a high concentration during temperature inversion smog conditions. In London between December 5th and 9th 1952, heavy continuous smog conditions caused an estimated 4000 deaths above the normal expectancy for December (Figure 6.1). The chief causes of death were bronchitis, pneumonia, and associated respiratory complaints. Similar smog conditions were also present in December 1962, but the number of deaths was only 700. The lower death rate 10 years later is attributed to the fact that the smoke concentration was lower, although the SO_2 concentration was

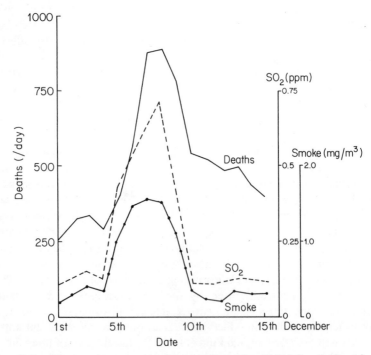

Figure 6.1. London smog conditions in December 1952 (after the Royal College of Physicians Report, 1970).

similar to 1952. It should be noted that the Clean Air Act was passed in 1956, and by 1962 smoke control orders made under the Act were in operation, so causing a reduction in smoke pollution. Similar smog conditions and an excessive mortality rate have been recorded in the Belgian Meuse Valley in 1930, and at Donora, Pittsburg in 1948. Smoke and SO_2 are most certainly one cause of bronchitis, and the condition is aggravated by polluted moist air conditions. Atmospheric smoke contains potentially carcinogenic organic compounds similar to those that occur in cigarette tobacco smoke, but no clear evidence has been produced to indicate that atmospheric pollution is a direct cause of lung cancer. It may however be an additive cause of the disease in heavy smokers. Other chest conditions that are aggravated by air pollutants are pneumonia, tuberculosis, and heart disease. Apart from these general effects on human health, there are other pollutants produced in specific industrial locations which cause disease. These are substances such as asbestos fibres, heavy metallic compounds, and numerous organic compounds.

6.3 Effects in the Stratosphere

The stratosphere has gaseous continuity and limited exchange with the troposphere below it. At present most of the air pollution occurs in the troposphere, and only comparatively small amounts of pollutants are found in the stratosphere. In the troposphere, the residence time for pollutants is restricted to a few weeks, and they are rapidly dispersed by air movements and removed from the troposphere in precipitated rain. In the stratosphere there is very little gaseous circulation and diffusion, so pollutants have long residence times, ranging from 1 to 3 years. Any transferred pollution from the troposphere to the stratosphere will tend to remain there and accumulate over a long period of time. In addition, there is a greater intensity of UV solar radiation in the stratosphere, and so more photochemical activity occurs involving smaller concentrations of pollutants than in the troposphere. Because of these differing stratospheric conditions there is concern about any build-up of pollutants in this layer.

The use of supersonic transport aircraft (SSTs) is a possible source of pollution in the stratosphere. These aircraft, such as the Anglo-French Concorde and military planes, fly at a height of about 20 km over non-populated areas of the globe. Also large numbers of subsonic civil aircraft fly in the upper troposphere, and some of their pollution could be transferred to the lower stratosphere. Jet aircraft engine exhausts emit relatively large quantities of water vapour and CO_2, and smaller amounts of CO, NO_x, SO_2, and hydrocarbons. Contrails or water vapour condensation trails are a familiar feature of the passage of jet aircraft in the troposphere. In the US, there is some evidence of an increase in the incidence of cirrus clouds along aircraft routes, paralleled by increasing jet aircraft activity. Contrails are also produced in the stratosphere, but they do not disperse to form clouds and can persist for up to eighteen months. This is because of the lack of air movement and the dry atmosphere. As more supersonic civil and military aircraft fly in the stratosphere in the future, then it seems possible that pollution may increase. The effects of this are not known but there are several possibilities that can be predicted. Increases of water vapour and CO_2 mean that a stratosphere greenhouse effect could be created, causing a rise in the earth's surface temperature, and changes in climatic patterns. Carbon monoxide and NO_x are pollutants which participate in photochemical reactions, and these involve the ozone layer present in the stratosphere. The US National Academy of Sciences in 1975 reported that 100 subsonic aircraft could reduce the stratosphere ozone by 0.02%, but 100 SSTs could reduce it by 0.7% and raise the earth's temperature by 0.5°C.

Another factor effecting the ozone concentration of the stratosphere may be the increasing use of pressurized aerosol spray cans. They are increasingly being used for products ranging from paint and pesticides to hairsprays and boot polish. In the UK, £200 M a year is spent on aerosol cans. The cans operate through the use of gaseous propellants of the freon type, which are chlorofluorocarbon compounds such as trichlorofluoromethane. These propellants are very inert and persistent compounds, which can be dispersed into the stratosphere as a pollutant, to the extent of between 10 and 100 000 tonnes annually. They may undergo photochemical changes, with the release of chlorine molecules which can dissociate ozone. There have been doubts expressed regarding the extent of the effects of fluorocarbons upon the ozone layer. However, one result of the aerosol spray controversy has been that manufacturers are attempting to find alternative non-halogen propellants, and the use of fluorocarbons has fallen sharply since 1974.

Any photochemical effect which reduces the ozone content of the atmosphere should be viewed with concern. The ozone layer reduces the amount of UV radiation reaching the earth, and any increase of this solar radiation would cause plant and crop damage and an increased incidence of skin cancer. However, recent research has shown that the chemistry of the ozone layer is not yet fully understood. Measurements indicate that the layer became 10% thicker over the period 1957–1970, but it then reduced at a rate of 3–6% between 1970 and 1973. These changes have been linked with the 11 year solar storm cycle and the intensity of solar radiation, rather than an increase in atmospheric pollution. Clearly more basic research into changes and conditions in the stratosphere is required, before the effects of increased pollution of the layer can be assessed. There is a need for more global monitoring of stratospheric constituents to supplement the four global monitoring sites that have operated since 1978. The short and long term effects of changes in the stratosphere do need to be fully understood in the future.

6.4 Biological Effects of Specific Pollutants

The ultimate aim of pollution abatement must be to reduce the amount of pollutants in the atmosphere to levels which do not affect the health of the population in both the short term and the long term periods. The short term period is within a person's lifetime, but the long term may extend over several generations, and needs standards to control substances that can produce genetic and inheritable effects. Environmental standards are described in the Department of the Environment Pollution Paper No. 11 as 'aiming to protect people, whether in their work place or in the environment at large, from hazards to health and safety that could result from products and substances consumed internally or released directly or indirectly into the environment'. There are generally two types of standards in use for places of employment and for the ambient environment (air and water). The work place standards are based upon exposure limits, or threshold limit values (TLVs), for specific substances, to which it is believed the majority of working adults may be exposed for a 40 hour week, and for a lifetime, without ill effects. The TLVs are determined from animal experimental data, medical knowledge and experience, epidemiology surveys, and environmental studies. The UK and other EEC countries usually use TLVs published by the American Conference of Government Industrial Hygienists (ACGIH) for workers in industrial environments (see Table 6.2). TLVs can be expressed in two forms: parts per million (ppm), defined as the ratio of the partial volume of the pollutant gas to the partial volume of air; and micrograms per cubic metre ($\mu g/m^3$), expressing the particulate loading of air, defined as the total mass of suspended particulate material per unit volume of the mixture. The setting of standards is far from easy. Many substances are potentially harmful to human health, but it is difficult to decide the point at which the concentration and the conditions required produce a risk to the health of the population. This is especially true where the harmful effects are cumulative in the human body, and long term effects result from exposure to low doses of the pollutant. Standards are determined from the best information available, which includes scientific and medical data, and the likely effects of long term exposure to individuals in their environment. The level decided must be set below that

Table 6.2. Threshold limit values for common pollutants (ACGIH)

Pollutant	ppm	mg/m^3
Carbon monoxide	50	55
Fluorine	1	2
Hydrogen chloride	5	7
Hydrogen fluoride	3	2
Hydrogen sulphide	10	15
Hydrocarbons	500	
Lead fumes and dusts		0.15
Nitrogen oxides	5	
Nitric monoxide	25	30
Ozone	0.05	0.2
Peroxyacetyl nitrate (PAN)	0.08	
Sulphuric acid		1
Sulphur dioxide	5	15

64

which causes detectable clinical symptoms, so that there is a safety margin included.

(12) The effects of the different Particulates in Atmospheric Pollution

Carbon Monoxide *(12) Bold*

This is a colourless, odourless gas produced during the incomplete combustion of fossil fuels, the refining of petroleum, wood pulp processing, and iron and steel smelting. The average global concentration in the atmosphere is 0.1–0.2 ppm and little is produced from natural biological processes. When breathed into the body, CO combines with the red blood pigment called haemoglobin, displacing oxygen, and carboxyhaemoglobin is formed. This results in less haemoglobin being available to carry oxygen round the body for tissue respiration and cellular energy production. People exposed to 80 ppm of CO have their blood-carrying capacity reduced by 15%, which is roughly equivalent to losing 1 pint of blood. The pollutant is a cumulative toxin and an acute concentration of 1000 ppm or more is invariably fatal. Increasing concentrations over 100 ppm cause headaches, dizziness, lassitude, nausea, vomiting, palpitation, breathing difficulty, muscular weakness, convulsions, and unconsciousness. The effects of low concentrations of CO over a long period are not fully known, and vary with age, sex, and the general state of health. It is suggested that heart and respiratory disorders are induced or exacerbated, but long term studies are complicated by other factors such as the effects of CO inhaled during cigarette smoking. In conditions of acute traffic congestion, CO levels as high as 400 ppm have been recorded for brief periods in the US, and toxic symptoms occur according to the duration of the exposure to the exhaust pollution. The CO concentration in urban air in the UK can reach 55–100 ppm, compared to the TLV for industrial workers of 50 ppm for an 8 hour day period.

Sulphur Dioxide

This gas has an unpleasant irritating odour and is produced from natural sources such as volcanoes, sulphur springs, and decaying organic matter, to the extent of 0.0002 ppm. The chief man-made sources of SO_2 are the combustion of fossil fuels, coke ovens, metal smelting, wood pulp production, petroleum refining, and brick manufacture. The global production of man-made SO_2 has been estimated at 100 M tonnes per year, and 90% is produced in the northern hemisphere. The effect of short term exposure to SO_2 is not very clear. However, some workers have been exposed to levels of 36 ppm for a number of years, and they have suffered respiratory symptoms such as naso-pharyngitis, coughing, and shortness of breath. Urban dwellers subjected to smoke, SO_2 pollution, and smogs also suffer from respiratory diseases like bronchitis, asthma and emphysema. Concentrations as low as 1–5 ppm can affect conditioned nerve reflexes and breathing, and if 10 ppm are inhaled for 10 minutes the pulse and breathing rates are increased.

Sulphur dioxide can be converted to dilute sulphuric acid in rain. Evidence that SO_2 is becoming an international pollutant is shown by increasing concern about the so-called acid rain that falls on some European countries and Canada. In 1958, rain falling over Europe had a pH of 5.0, but by 1962 it was 4.5 in some countries such as the Netherlands. Sweden experienced a fall in the rain pH to 4.5 over the period 1962 to 1966. This acid rain caused leaf damage and a reduced growth rate in Swedish forests, which are an important natural resource for the production of wood pulp, paper and board. In the early 1970s further ecological damage was found, and by 1979 it was estimated that 20% of Sweden's lakes, or 20 000, were affected to the extent that they had reduced numbers of fish, or none at all. Similar pollution effects are present in rivers and lakes in south-western Norway. The pollution is thought to be carried to Scandinavia on the prevailing south-west winds from the UK, the Ruhr, and Germany. Sulphuric acid and particles of cadmium and lead are deposited on the winter snows, and when these melt the pollutants enter the rivers and lakes. The contaminated melt water enters water courses at the time when fish spawning and hatching is occurring, and this kills the eggs and young fry, so polluting the ecosystems. In 1979, it was estimated that the sulphur emission rate for Europe was up to 70 M tonnes per annum. Atmospheric pollution by SO_2 must be reduced in the future, perhaps by using some of the methods described in Chapter 5.2.

Ozone

This gas occurs naturally at low levels of the atmosphere to the extent of 0.02 ppm, and it is increased by photochemical smog caused by excessive vehicle exhaust pollution. At high concentrations, between 1.5 and 2.0 ppm, ozone can cause temporary effects within 2 hours, producing irritation of the eyes, throat and lungs, and retarded mental capacity. People prone to asthma attacks can be adversely affected by concentrations as low as 0.25 ppm. There is uncertainty about the longer term effects of the gas, and apart from aggravating respiratory complaints it may act as a tumour accelerator. Surveys of ozone in the atmosphere over Los Angeles have recorded peak levels of 0.9 ppm, compared to the highest peak level over the London area of 0.1 ppm in exceptional heatwave conditions.

Nitrogen Dioxide

This is a reddish-brown gas with a pungent odour and it is highly toxic. It is the main NO_x type of pollutant, and is produced from vehicle exhaust systems and some chemical manufacturing processes. Like ozone, it can be formed by photochemical action and so is present in photochemical smog. The gas when inhaled can form nitrous and nitric acids which attack the mucous inner lining of the lungs. The acute symptoms are pulmonary fibrosis (scarring) and emphysema (distension) of the lungs, which has occurred in workers who have been regularly exposed to levels of 10–40 ppm of NO_2. The effects of lower concentrations are not known, beyond irritation of the eyes, throat and lungs.

Lead

A number of heavy metals are toxic to the human body in quite small concentrations, and lead is no exception. The use of lead pipes for our water supply over generations has meant that lead is present in water and food. It is also present in the soil, and in rivers and reservoirs supplying drinking water. To this natural lead content must be added the lead from petrol vehicle exhausts, pesticides, and the burning of coal and industrial rubbish. Lead is taken into the body via the lungs and the alimentary canal. Excessive lead in the body can cause weakness, lack of muscular co-ordination, miscarriage, anaemia, and damage to the nervous system and the kidneys. Damage to the brain (lead encephalopathy) probably causes mental retardation in young children, who seem to be more susceptible to lead poisoning than adults. In addition, lead is a cumulative toxin in the body where it replaces calcium in bones. The World Health Organization (WHO) limit for lead in drinking water is 0.05 ppm. Maximum blood lead levels have been set at 0.7 ppm, but many medical authorities consider the maximum blood levels should be 0.4 ppm for adults, and 0.3 ppm for children. The normal blood lead level is taken as 0.2–0.3 ppm derived from food and drink intake, but this level has increased sharply over the last 30 years in the general population. The increase has been ascribed to airborne lead, especially the proportion derived from petrol. In GB, it is estimated that about 10 000 tonnes of lead per year enter the atmosphere from vehicle exhausts. This pollution is not confined to urban areas alone but is international in its occurrence. Environmentalists are concerned that the lead present in remote areas such as the Greenland icecap, showed a 300% increase between 1940 and 1965. Although the global atmospheric lead content is increasing rapidly, there is much more in city air. In some US cities, at periods of great traffic congestion, a lead level as high as 25 μg/m^3 has been recorded, compared to a level of less than 1 μg/m^3 in rural areas. Garage mechanics in the US may have 0.34–0.38 ppm of lead in their blood compared to the recommended maximum blood level of 0.55 ppm for industrial workers. In the UK, there is certainly evidence of increased levels of lead in children and adults who work or live near motorways and interchanges such as Gravelly Hill ('Spaghetti Junction') near Birmingham.

Particulates

Solid particles of various diameters are emitted into the atmosphere from a variety of sources. These include mineral dusts, fibres, and a range of manufactured chemical compounds. Mineral dusts are coal, stone, fireclay, mica, china clay and graphite; and fibres include asbestos, glass fibre, and rock wool, which are used for insulation purposes. Inhalation of these particles irritates the lungs, and over a period of time can cause scarring or fibrosis of the lung lining. Pneumoconiosis or dust disease, silicosis or stone dust disease, and asbestosis are well known industrial pollution diseases. Prolonged exposure to asbestos, and possibly glass fibres, over a period of twenty years or longer is known to cause mesothelioma, or cancer of the lung lining or abdomen. Asbestosis was first recognized in 1906, and by 1930 a connection was established between asbestos and lung cancer, and with gastric cancers by 1964.

Asbestos is widely used for building construction, heat insulation, pipe lagging, and in gaskets and brake linings for vehicles. The use of asbestos causes microscopic fibres of magnesium silicates to separate from the materials into the atmosphere. There are two main types called blue (crocidolite), and white (chrysotile) asbestos. Although some UK authorities consider that blue asbestos is more dangerous than the white form, it is very likely that both forms are equally hazardous. Indeed, it has been stated that even a short exposure of 6 months to a heavy dust concentration can eventually cause asbestosis. Past neglect of the dangers of using asbestos caused 800 recorded industrial fatalities up to 1976. The thermal insulation industry banned the use of asbestos in 1968, and there has been a voluntary ban on the use of blue asbestos in the UK since 1970. In 1969, new Asbestos Regulations stipulated stringent precautions for people working with asbestos. Safety standards of 2 fibres/cm^3 of air for white asbestos, and 0.2 fibres/cm^3 for the blue form, are in operation. Some 70 years after asbestosis was first known, the Government set up an Asbestos Advisory Committee to investigate in 1976. The Committee reported in 1979, and recommended a new safety standard of 1 fibre/cm^3 of air for white asbestos, which is still being used. The continued use of asbestos should be questioned, and efforts are being made to use plastic foams for thermal insulation, and polypropylene in cement for sheeting and pipes. In relation to other fibres, a Health and Safety Executive report in 1979 recommended safety limits for nuisance dusts of 10 mg/m^3, and 5 fibres/cm^3 for man-made fibres.

6.5 Radiation Effects

When radiation passes into and through the human body, energy is transferred to the biological atoms and molecules that make up the tissues. This causes ionization or loss of electrons from the constituent atoms. In theory, each single electron that is set free can cause damage. The probability of this happening is somewhat remote, but a series of ionizations can cause disruption of molecules, and upset the balanced functioning of biological systems. Radiation damage to the body is not often detectable immediately after an exposure to radiation, or for some years later. This may be because the natural biological repair mechanism operates, with the result that no apparent damage is seen. However, it is believed that initial radiation damage can cause further disruptive effects, that over a long period of time produce serious damage to tissues and organs. This is not fully understood, but it is known that long term radiation damage tends to be selective, so that some parts of the body are more sensitive to damage than others. The radiation damage produced is proportional to the type of radiation, the energy level, and the time and frequency of the exposure to it. Very high doses of radiation can produce death, either immediately, or after a few days. Lower doses can cause no immediate symptoms but within a few days 'radiation sickness' occurs. This can eventually be fatal, but if recovery takes place there is usually permanent impairment of health. The biological effects can be classified as somatic or genetic. Somatic or body effects show within an individual's lifetime, and appear as skin damage, eye cataracts, damage to liver, spleen and thyroid, and reduced fertility. Other effects are leukaemia, where the red bone marrow produces uncontrolled numbers of white blood cells, and various forms of carcinoma where there is uncontrolled growth of cancerous cells in the body. Genetic radiation damage is not apparent in one generation, but can appear in offspring and in subsequent generations. Radiation affects the parental reproductive organs, and so the gametes or sex cells produced contain mutations that are inherited. Within these structures, the genetic mechanism of the chromosomes containing DNA becomes damaged by ionization and breakage. Medical studies carried out on the Japanese who survived the atomic bomb explosions at Hiroshima and Nagasaki in August 1945, and their offspring, have confirmed the effects of radiation damage. Also in 1976, a survey was carried out in India of 13 000 people who lived in an area with a high thorium-234 soil content. It was found that the incidence of severe mental retardation in children was 3 in 1000 of the survey sample, compared to 1 in 1000 in a neighbouring area of normal background radiation. Chromosome damage was found, and a higher than normal number of cases of mongolism were present. This effect is paralleled by the evidence that mothers who have received abnormal doses of X-rays for diagnostic purposes during their pregnancy tend to produce an above average number of mongol children. There is no doubt that all types of radiation, including X-rays, can be harmful to life. It has been stated that X-rays probably cause between 2500 and 100 000 genetically defective births, and 200 to 300 leukaemia deaths annually. In perspective, these figures represent a tiny fraction of the world population, and excluding X-ray effects, there are about one million genetically defective births each year.

It is difficult, if not impossible, to decide the level of radiation that can be assumed to be completely harmless. The maximum permissible dose (MPD) of ionizing radiation was originally considered to be 'one that was not expected to cause appreciable bodily injury during the lifetime of any occupationally exposed person' (International Commission on Radiological Protection (ICRP) 1966). Appreciable bodily injury was considered as any effect regarded as deleterious to health. The ICRP believes that the MPD is not applicable to the population at large, and so they recommend dose limits (DLs) in this context. The evidence that can be used to determine the MPD and DLs values is relatively small. Some evidence is available from studying the effects of X-rays and medical isotope therapy, as well as the survivors of the two atomic bomb explosions in the Japanese cities of Hiroshima and Nagasaki in August 1945. Other information is available from reports of the effects of laboratory accidents, or the accidental release of highly radioactive fall-out. In all these cases, the body usually receives massive doses of radiation that are unlikely to be received by the majority of occupational workers or the general public. The evidence that is most needed is that relating to the long term effects of cumulative small doses of radiation.

Animal irradiation experiments show that cancers and leukaemia are likely effects, and there is also a reduction of the normal life span. American studies on radiologists in 1957 showed that their average life span was 60.5 years, compared to the average of 65.6 years for the US population generally.

There are clearly considerable difficulties about determining MPDs and LDs. It is not possible to directly relate animal irradiation dosages and their effects to human beings. It is not known why cancers resulting from radiation exposure only occur after a long delay of years. It is very difficult to assess the genetic effects of radiation because of the time span of generations, the masking of recessive gene effects, and the problem of separating natural from induced mutations. It is now generally agreed that any dose of ionizing radiation, no matter how small, must be regarded as potentially harmful. In the face of these and other difficulties, the assessment of MPDs and LDs can only be made by using the best expert opinion available, and they must be under constant review. Like any other pollution effect the biological costs of radiation must also be balanced against the beneficial effects of radiation therapy and diagnosis, and the economic benefits of increased energy production. The ICRP therefore recommend doses that are regarded as tolerable, if sufficient benefits are available to outweigh the risks involved. The MPD recommendations have been successively reduced from 7280 mrem per year in 1931, to 500 mrem per year in 1957. The 1966 ICRP recommendation for occupational workers is 5000 mrem of whole body radiation per year, and various higher doses for specific organs and exposed parts of the body. The DL recommended for members of the general public is a maximum of 500 mrem per year, which is one tenth of that for occupational workers who are subject to strict monitoring procedures. The ICRP recommended genetic dose limit is a maximum of 5000 mrem from all sources, in addition to the radiation from the natural background and medical procedures. This genetic dose is based upon a mean child-bearing period of 30 years, and is equivalent to 150 mrem annually. The Maryland Academy of Sciences in the US has calculated that the genetic handicap risk of 170 mrem per year for a birth rate of 4.6 M annually is 2 per 1000 births. Overall it appears that the gross annual radiation dosage to the population at large should not exceed about 150, plus 130 (background), plus 20 (medical), which is equal to 200 mrem.

6.6 Pollution Accidents—A Case Study

Accidental discharges of toxic or hazardous pollutants occur from time to time. These isolated incidents do not produce a general pollution effect, and only affect a relatively small localized population. However, a close study of the causes and effects contributes valuable information regarding what happens when there is a high concentration of atmospheric pollution.

One such isolated incident occurred on July 10th 1976 at Seveso, just north of Milan in Italy (see Figure 6.2). An explosion took place in a chemical plant which manufactured trichlorophenyl. The result was that a white cloud of poisonous gas was released, containing dioxin or TCDD. This toxin was a by-product of the

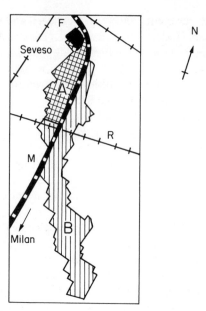

Figure 6.2. The Seveso district: F, factory; R, Railway; M, motorway. For explanation of zones A and B, see text.

production of the herbicide 2,4,5–T, and TCDD is extremely poisonous to humans. The gas cloud settled over the nearby town of Seveso, and contaminated the buildings and the ground. Soil samples were taken at less than 100 sampling points, chosen on the assumption that the wind was blowing from the north-east on the day of the accident.

Eventually an area of 1430 hectares of heavy contamination was designated as zone A (Figure 6.2). The actual area designated was almost certainly incorrect for a number of reasons. It is probable that very low wind conditions prevailed on July 10th, with the result that the poisonous gas moved haphazardly rather in a fixed direction. This is supported by the evidence of a random pattern of deaths to small animals, to the west, north, and east of zone A in the weeks following the accident. The zone A boundary also had certain peculiarities of shape. It zigzagged to exclude some industrial and prosperous housing districts, included the Como to Milan motorway along the eastern side, but the southern boundary stopped short only 90 m away from the Milan to Switzerland and Germany railway line. Soon after the explosion, the inhabitants in zone A and outside it developed skin disorders consistent with dioxin contamination. After an inexcusable delay of 3 weeks, the Italian Government decided to compulsorily evacuate the 780 inhabitants of zone A, but the motorway and the railway continued to operate normally. At a later date a further area, called zone B, was designated, but there was no evacuation of the 6000 inhabitants, because the contamination was considered to be 'tolerable'.

The original dioxin pollution continued to spread, and eventually covered a much greater area than both zones A and B. Immediately after the accident the pollution spread was in the atmosphere, and particulates were deposited over an unknown area. Following this, there was obviously a further spread of the ground-deposited contamination by various means. Transport on the motorway and the railway was one of the obvious methods, and another was the behaviour of the owners of property, and the general public. People were able to enter zone A through lax security, either to rescue their belongings or to carry out property and loot. Vegetables growing only 8 km south of Seveso, although outside the designated zone A, were harvested and sold in local markets. In the autumn of 1976, the River Seveso went into flood three times, and the run-off from the land spread contamination along the course of the river and into Milan. These occurrences ably illustrate the manner in which pollution can spread in the environment through natural behaviour.

There have been considerable health effects as a result of the dioxin pollution. Skin chloracne, which showed as boils and pimples, affected 187 people including many children. Soon after the accident, 46 other people suffered skin and liver complaints. In addition 32 abortions have been carried out with official approval, and probably up to another 80 women have aborted unofficially. In 1976 there were 2959 births in zones A and B, and amongst the offspring, four were deformed, and thirteen were born prematurely. Between January and May 1977, among the 1141 births there were eleven deformed and six premature offspring. These figures are above the norm for this district of northern Italy. So far no deaths have taken place, but the long term effects of dioxin toxicity are totally unknown.

Experts consider that further spread of dioxin will continue on the ground surface, and also through the movement of soil water. Drainage into rivers such as the Seveso, which runs into the Po, could eventually effect water resources and the natural biological cycles. A conference of international experts on dioxin produced conclusions of considerable concern. They considered that probably an effective tolerance level for humans did not exist. Certainly it seems that tolerance levels adopted by the Italian Government, Ministry of Health for the designation of the zones were too high, perhaps by a factor of 1000. The history of this serious pollution accident emphasizes a number of fundamental points about control methods. The assessment of the meterological conditions, the sampling methods, and the determination of correct tolerance values are obviously of crucial importance. These should be the basis upon which contaminated zones are designated. Once designated, then strict precautions must be taken within the zones in respect of entry by the public, the movement of transport in and out, and the movement of produce and goods out of the zones. These are essential to prevent the spread of ground pollution to non-contaminated areas. Considering the broader aspects of accidental pollution, then there is a need for thorough research and monitoring of all known toxic pollutants that are manufactured. The necessary emergency pollution prevention plans must be drawn up for specific toxins, and used to limit spread, and minimize health damage. This appears to be the price required by the advance of science and technology.

CHAPTER 7

Future Developments

Present atmospheric pollution originates from two main sources, namely the inefficient combustion of primary energy sources and the emission of waste industrial and domestic products into the air. There are broadly three possible changes that could reduce the level of pollution: first, if the usage of coal, oil, and gas were to be reduced; second, if the process of combustion were to become more efficient; and third, if less waste products were discharged into the atmosphere.

The recent trends of fuel utilization have shown a reduced use of coal and an increased consumption of oil and natural gas. The development of North Sea oil and gas reserves will encourage this trend until the end of this century, but in the next century it is probable that oil and gas supplies will become scarcer. Consequently, it is expected that the use of coal and nuclear fuels will increase, to provide the main energy sources for the future. To these may be added newly developed renewable natural resources such as wind, solar and tidal power. These changes in the source and utilization of fuel resources could produce a reduction of atmospheric pollution. The renewable energy sources are all non-pollutive, and over two decades it is possible that less pollutive techniques for the combustion of coal could be developed. Nuclear power is not likely to greatly pollute the atmosphere, but it could pollute the environment if fuel and waste recycling and radioactive waste disposal problems are not solved satisfactorily. The other consideration is the future energy demand. The present economic policies of the Developing countries are committed to expansion, increased manufacturing production, and increased exporting of goods and services. Expansion opportunities are envisaged in the future, particularly in relation to supplying those Developing and Third World countries who wish to develop their national economies with the revenues from their natural resources. If this hoped-for expansion occurs, then energy fuel consumption and production will inevitably rise. This could be offset, to some extent, by energy conservation measures, but these would have to be on a considerable scale to significantly reduce future energy demand. Expanding industrial production will continue to increase waste production and may cause additional pollution. The present policy of discharging waste into the atmosphere and hoping that it will be sufficiently diluted to be innocuous, cannot be allowed to continue in the future. Critical atmospheric concentrations will occur more frequently if this practice continues. Hazardous pollutants will have to be removed from waste, or rendered innocuous at source, by the producer.

Long term pollution trends in the future are uncertain and the subject of argument, but in the short term antipollution operations and measures can be improved. The starting point for improvement should be regular monitoring of the atmosphere for hazardous pollutants using a national network of sampling stations. Complementary to this is the need for improved knowledge of chemical reactions in the atmosphere, and particularly in the stratosphere. The long term effects of small cumulative doses of pollutants upon occupational workers and the general public also needs more research. The improved knowledge of atmospheric concentrations and the human effects of specific pollutants will enable satisfactory assessments of threshold limit values to be made in order to safeguard the population. The problems of pollution reduction can then be examined in the knowledge of what has to be done to meet the required TLVs. The reduction of the pollution of the atmosphere to acceptable levels requires both national and international cooperation and action. It is encouraging to note that EEC members are studying pollution problems, but this cooperation should also include the Scandinavian and Baltic Sea countries. Uniform international standards and agreement upon common pollution abatement measures may well be important future objectives. In practice they will probably be difficult to achieve for several reasons. A widespread system of monitoring and enforcement would be required, and the amount of pollution abatement required would vary from country to country. Also the considerable cost involved would not necessarily be proportional to the size of each national exchequer, and the availability of finance would be subject to varying national priorities.

References

Boughey, A. S., *Man and the Environment*, 2nd edn., Macmillan, 1975.
Clayton, K. M., and Chilver, R. C., *Pollution Abatement,* David and Charles, 1973.
Coppock, T. T., and Wilson, C. B., *Environmental Quality*, Scottish Academic Press, 1974.
Crawford, M., *Air Pollution Control Theory*, McGraw-Hill, 1976.
Department of the Environment, *Clean Air Today*, HMSO, 1974.
Goldsmith, E., *Can Britain Survive?*, Tom Stacey Ltd, 1971.
Higgins and Burns, *Chemistry and Microbiology of Pollution*, Academic Press, 1975.
Holister, G., and Porteous, A., *The Environment: A Dictionary of the World Around Us*, Arrow, 1976.
Johnson and Steere, *The Environmental Challenge*, Holt Rinehart, 1975.
Massachusetts Institute of Technology, *Man's Impact on the Global Environment*, MIT Press, 1971.
National Society for Clean Air, *Clean Air Year Books*, NSCA.
National Economic Development Office, *Energy Conservation—CPRS Study*, HMSO, 1974.
Open University, *Maintaining the Environment, T100 Units 26 and 27*, O. U. Press, 1972.
Patterson, W. C., *Nuclear Power,* Penguin, 1976.
Rothman, H., *Murderous Providence*, Rupert Hart-Davis, 1972.
Royal Commission on Environmental Pollution, *First Report*, HMSO, 1971.
Royal Commission on Environmental Pollution, *Second Report*, HMSO, 1972.
Royal Commission on Environmental Pollution, *Third Report*, HMSO, 1972.
Royal Commission on Environmental Pollution, *Fourth Report*, HMSO, 1974.
Royal Commission on Environmental Pollution, *Fifth Report*, HMSO, 1976.
Saunders, P. J. W., *The Estimation of Pollution Damage*, Manchester University Press, 1976.
Scorer, R. S., *Pollution in the Air*, Routledge and Kegan Paul, 1973.
Smith, K., *Principles of Applied Climatology*, McGraw-Hill, 1975.
Tearle, K., *Industrial Pollution Control*, Business Books, 1973.
Tucker, A., *The Toxic Metals,* Earth Island Ltd, 1972.
UK Atomic Energy Authority, *Evidence to the Royal Commission on Environmental Pollution*, HMSO, 1976.
Walker, C., *Environmental Pollution by Chemicals*, Hutchinson, 1971.
World Health Organisation, *Health Hazards of the Human Environment*, WHO, 1972.
Yapp, W. B., *Production, Pollution, Protection*, Wykeham Publications, 1972.

SECTION III

Land

CHAPTER 8

Pollution and the Use of Land

The term land pollution does not have a very precise meaning. This is because there are different types and degrees of pollution, and people have varying standards by which they judge pollution effects. Land pollution may be regarded as the despoilation of the urban and rural scene. In rural areas many people appreciate and enjoy the countryside for its natural state, unspoilt by man's activities. Consequently any industrial operations that disrupt the natural landscape surface are considered to be pollutive. Similarly any human activity that adversely affects plant and animal wild life is also pollution in the view of nature conservationists. Dereliction is another form of pollution, whether it occurs in cities or in the countryside. It is regarded as an eyesore because it presents an untidy, neglected environment, which is unpleasant and unsuitable for agricultural, or industrial, or residential purposes. Alternatively these views on land pollution must be seen in perspective alongside the realities of our highly technological age. We must continue to extract fossil energy and mineral resources from the earth, so long as the present economic growth policies are pursued. Continued extraction inevitably means that the waste produced must be disposed of. To sustain the present level of food consumption in the UK the agricultural industry must seek to increase home food output and improve production efficiency. This means changes in farming practices and the continued use of agricultural chemicals. Also industrial and economic survival in a highly competitive import–export world means that new technologies must be developed, and production methods must change. Consequently, obsolescence and dereliction of plant and equipment will continuously occur. Therefore so long as the national economic growth policy continues, accompanied by increased industrial output and increased services and standards of living, then continuing land pollution seems inevitable. Many people accept that land pollution is a necessity, but there is growing concern about the amount of pollution that can be contained within the environment, both now and in the future. There is an increasing consciousness of the need to maintain and conserve the aesthetic quality of the environment. There are changing standards about what is pollution, and should not be tolerated. The view of environmentalists is that land pollution must be reduced and better contained in the future, otherwise it will go on increasing and reach an intolerable level.

A more objective definition of land pollution is 'any physical or chemical alteration to land which causes its use to change and renders it incapable of beneficial use without treatment'. Alternatively pollution can be stated as misuse of land, disuse of land, and chemical contamination of land. Four main types of land pollution can be distinguished:

1. Land that is affected by the active tipping of solid and liquid waste material. Tips and spoil banks are usually the result of mining and quarrying operations, or the need to deposit domestic, trade and industrial waste. If the tips and dumps are no longer in use, and are not covered by natural vegetation, then they may be described as derelict.

2. Land that is derelict. This can be classified into four categories:

    2.1. Disused land that has dilapidated or abandoned buildings and equipment on it. This is often found in industrial urban areas, but can also occur in rural districts, for example where abandoned wartime aerodromes and defence sites are located.

    2.2. Disused land that was once used for mining and quarrying, but the operations have now closed down. This is characterized by excavations and spoil heaps.

    2.3. Disused land that was formerly used for rail passenger and freight, or canal transport. This type of dereliction increased in the late 1960's when national policy decisions caused the abandonment of uneconomic transport routes.

    2.4. Land that is not under active cultivation, maintenance or conservation. This type of land may be called neglected, or under-utilized, rather than derelict. It is found in scattered areas throughout the UK, and is often used for the indiscriminate dumping of waste.

3. Land that is used for the disposal of industrial wastes containing hazardous chemical and radioactive substances. This results in the land being unsuitable for farm cultivation or human contact.

4. Agricultural land in active use and where the husbandry methods are likely to cause water pollution. The use of potentially toxic chemicals on crops and the land affects the drainage water entering streams and rivers.

The Use of Land

Land in any country is required by man for a number of uses. These may be broadly categorized as: agricultural and horticultural food production; afforestation; housing, commerce and the service industries in relation to urban and rural communities; industrial sites; transportation by road, rail, water and air; and for recreation and amenity purposes. In the UK there are numerous problems related to land utilization. In general terms there is a shortage of land in the industrial and densely populated urban areas, for example the midlands and south-east, and a surplus of land in the rural areas such as the Scottish highlands, north and south-west England, and north and mid-Wales. An analysis of the total UK land surface of about 24.1 M hectares, made in 1976, is given in Table 8.1. Although there appears to be 19.14 M ha of land available for agricultural

Table 8.1. Land Use in the UK, 1976

Arable and grassland	12.06M hectares	50.0%
Rough grazing and hill land	7.08	29.4
Forest and woodland	1.78	7.4
Urban land	1.85	7.7
Other land	1.33	5.5
Total	24.1	

use, in fact only about 12.3 M ha or 51% were used for food and grass production in 1976. The balance of 6.84M ha was either unsuitable, or uneconomic for use because it was rough grazing or hill land. The 51% of farmed land only produced 53.7% of the UK food requirements in 1977, so that 46.3% had to be imported. Therefore there is insufficient suitable land for food production, and the nation is only about half self-sufficient in total food requirements. Also the land in agricultural use is contracting each year, to the extent of about 415 000 ha or 3.3% each decade in the UK. This land is removed from food production because it is used for the construction of motorways, road improvement schemes, new housing and industrial developments, increased tipping of waste, etc. The urban population also require the use of land for recreational and sporting purposes, and this will become more important in the future as the working week is shortened. About 1.66 M ha of land in England and Wales is safeguarded from industrial and urban development, and is designated for the public's enjoyment. This consists of ten National Parks of 1.362 M ha under the Countryside Commission, 110 000 ha owned or leased by the Nature Conservancy, and 19 000 ha owned or leased by the National Trust. Nature conservationists concerned with the preservation of wild life, and the availability of natural land for recreational purposes, are continually pressing for an increase in the area of this protected land. The UK water demand is also increasing at about 5.5% per annum, and part of the water supply is drawn from water storage reservoirs which utilize land. One way to meet this increasing water demand is to construct more reservoirs, but this entails using more land which could be available for agricultural or recreational purposes. About 7.4% of land is used for forestry and there is an increasing demand for more timber and paper. In 1977, UK forests only provided 8% of our requirements, and so there are forward plans to double home production of timber by the year 2000, which will again require more land. The disposal of waste utilizes land, and the amount of solid waste produced is rising each year for several reasons. More consumer goods are being designed and manufactured with shorter working lives. The public are dissuaded against the repair and retention of goods, and encouraged to replace and discard worn-out and superseded products. This built-in obsolescence and replacement policy for consumer goods, aided by competition, sustains an expanding manufacturing demand. This in turn requires new plant and equipment, and a more rapid turnover of materials and goods that inevitably produces more waste. There is also continual modification and re-equipping of houses, shops, and offices, as well as the demolition and construction of new buildings. All these developments, which are an accepted part of modern life in this country, contribute to the production of more and more waste that requires land surface for its disposal.

This brief account of some of the problems of land utilization in the UK clearly shows that there is an overall shortage of land for all the various uses. In addition there is conflict and competition between the various land requirements and users. Land pollution in terms of waste deposition and dereliction is a misuse of land. If this increases, less land will be available for other essential and more useful purposes. There is a pressing need to make maximum and effective use of all areas of UK land, both now and in the future.

CHAPTER 9

Waste Disposal and Land Pollution

9.1 The types and Quantities of Solid Waste

Waste can be classified in different ways, but for present purposes the types of waste will be considered to be domestic, trade, and industrial, relating broadly to their origins. There are no precise definitions of these categories, but domestic or household waste usually means refuse from houses and other residential premises. Trade waste is refuse arising from retail, commercial and business premises; and industrial waste arises from all mineral, manufacturing, and processing sites and premises. Industrial waste can be conveniently sub-divided into waste from coal mining and the mineral extraction sector, the manufacturing and processing sector, and the nuclear power industrial sector.

The quantity of waste produced in the UK is of the order of 165 M tonnes per

Table 9.1. Estimates of annual solid waste production

Type of waste	Quantity (M tonnes)	Percentage
(a) Received by Local Authorities 1966–67		
Domestic and trade	14.0	71.8
Industrial	1.9	9.7
Construction industry	3.3	16.9
Other types	0.3	1.5
	19.5	11.85%
(b) Produced by industry 1973		
Coal mining spoil	58.0	40.0
China clay quarrying	22.0	15.2
Other quarrying	27.0	18.6
CEGB power stations	12.0	8.3
Mining other than coal	3.0	2.1
Other industries	23.0	15.9
	145.0	88.15%
Overall total	164.5 M	tonnes

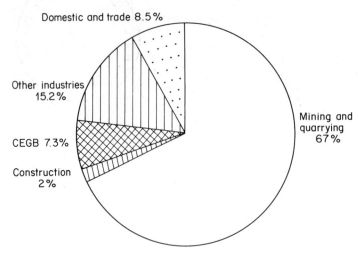

Figure 9.1. Annual solid waste production.

annum. There are no accurate figures available, but Table 9.1 provides some indication of the size of the problem of waste disposal. Table 9.1 (*a*) shows the waste received and disposed of by Local Authorities in 1966–67. They only disposed of about 12% of the total waste produced, and this was derived mainly from domestic and trade premises. By comparison, Table 9.1 (*b*) shows the solid waste produced by industry in 1973. About 76% or 110 M tonnes of this industrial waste was the result of coal mining, mineral working, and quarrying operations. The Central Electricity Generating Board (CEGB) power stations produced 8.3% or 12 M tonnes of pulverized fuel ash. The remainder of the industrial sector only produced 11 M tonnes, but this was of very varying composition. Figure 9.1 shows the total waste production. As far as industrial waste disposal is concerned, then the extractive industries present a quantity problem, and other industries create a quality problem, in terms of solid and slurry waste containing potentially hazardous chemical compounds. This type of waste is usually described as 'toxic'. The Department of the Environment Technical Committee on the Disposal of Toxic Wastes Report of 1970,

Table 9.2. TCDTW survey of industrial solid and semi-solid wastes, 1970

Type of waste	M tonnes/yr	Percentage
General factory waste	1.18	10.5
Relatively inert process waste	9.32	82.6
Flammable process waste	0.13	1.2
Acid and caustic waste	0.44	3.7
Indisputably toxic waste	0.21	1.8
Total	11.28	

Table 9.3.   Some types of toxic wastes

The Technical Committee on the Disposal of Toxic Wastes Report 1970 included a list of the types of wastes mentioned in the replies to their survey. This is one means of illustrating the meaning of the word 'toxic'.

Tarry liquids
Solid tarry matter
Sludge from tar distillation
Acid tars
Oil impregnated rubbish
Waste oil
Lubricants
Water–kerosine mixtures
Waste paint
White spirit
Lacquers
Spent sheep dip
Residues from pesticide products
Mercaptans
Photographic wastes
Nicotine wastes
Kier liquor
Plating sludges
Pickling sludges
Spent acids
Leaded petrol sludges
Sludges containing copper, zinc, cadmium, and nickel compounds
Arsenic wastes and arsenious sulphide
Beryllium wastes
Cyanide wastes
Alcohol wastes
Sulphides
Fluorides
Alkaloid wastes
Aromatic hydrocarbons
Complex cyanides
Chlorphenols
Chlorcresols
Carbides
Chrome acid
Trichloroethylene
Beta-naphthylamine sulphate
Diaminodiphenylmethane
Propyl isocyanate
Sodium acetylide
Highly acid organic residues
Noxious organic solvents

The list of wastes in Table 9.3 may be broadly classified into: acids, alkalis, metals and metallic compounds, aqueous organic, aqueous inorganic, and non-aqueous materials.

(TCDTW) did not clearly define toxic wastes. The Committee organised a survey of 1186 firms that were thought to produce some toxic wastes, but the firms were left to classify the wastes in their own way (Table 9.2). In Table 9.2, it appears that the only toxic waste shown is the 1.8% designated as indisputably toxic waste. Generally the word toxic means poisonous, or it refers to a substance which when introduced into, or absorbed by a living organism destroys life or impairs health, over a short or long term period. But as the TCDTW report states, every substance can be toxic if the concentration and the quantities absorbed are sufficiently great. Toxic in relation to wastes therefore appears to depend upon the quantity present within the total amount of waste. For example a few kilograms of phenol in 10 000 tonnes of waste might not be considered to be toxic. Environmentally, the term toxic pollutant depends upon the quantity entering the environment and individual organisms, and the effect produced. Therefore if all the phenol mentioned above eventually enters a small water-course it will cause toxic pollution. In Table 9.2 it should be appreciated that only 1.8% was designated as indisputably toxic. But some or all of the flammable, acid and caustic waste was probably also toxic, so that the total toxic waste could have been as high as 14%. Table 9.3 provides some guide to this view. In 1974, about 3 M tonnes of toxic waste from all industries was notified for disposal on land. This was broken down into 25% flammable, 25% chemically toxic, and 20% acid toxic. Another estimate is that about 10.8 M tonnes of toxic waste is produced annually, but probably 8.6 M tonnes or 80% of this is environmentally inert. Therefore about 2.2 M tonnes is potentially hazardous. Although the exact amount of toxic waste produced in the UK is not known, within the total waste output the probable amount is relatively small at less than 1.5%.

9.2 Methods of Waste Disposal

It is inevitable that as there are different types of waste, there will be varying methods of waste disposal. Briefly most solid wastes are deposited on land as tips or spoil heaps, or as land infill to quarries and mine shafts, or as dumps containing a large range of materials. In addition small quantities of waste are dumped into the sea. The methods of disposal are summarized in Figure 9.2. The diagram shows that there are various stages to the process. Waste is produced continually so there is often a need for some sort of storage facility. In the case of some mineral extractive industries such as deep mined coal, china clay and ironstone, there is storage on the working site as spoil heaps, but this is waste deposition rather than disposal. In other industries the stored waste often has to be transported to disposal areas and tipped or dumped. Alternatively, the stored waste may be treated in various ways before disposal. The treatment may reduce the bulk, or make disposal easier, or extract materials that can be re-used or recycled back into manufacturing processes. In respect of environmental pollution the quantity, the treatment, and disposal methods of waste are of prime importance.

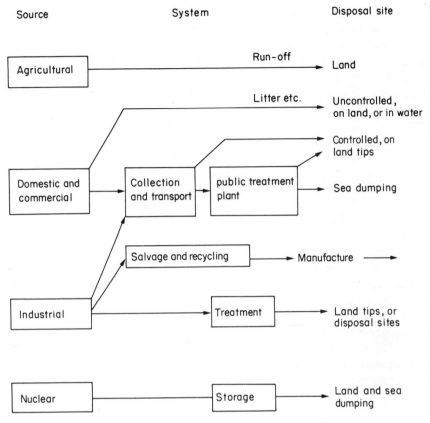

Figure 9.2. Outline of solid waste disposal.

Methods Used by Local Authorities

The Public Health Act 1936 enables Local Authorities to collect, treat, and dispose of all refuse from the domestic sector, and such industrial and trade waste as requested. In 1973, Local Authorities in England dealt with 19.5 M tonnes of waste by various methods. (see Table 9.4). About 15 M tonnes of the tipped waste is household refuse consisting of cinders, ash, dust, vegetable and waste

Table 9.4.   Disposal of Household and Trade Waste, England and Wales 1972–3

Method	Quantity in M tonnes	Percentage
Tipping	16.8	86.3
Incineration	1.85	9.5
Pulverisation	0.7	3.7
Composting	0.095	0.5

food matter, paper, board, metal, rags, glass, and plastics, whilst the remaining 4.5 M tonnes is from trade sources. About 86% of this waste is not pretreated and is disposed of by land tipping. One quarter of this waste is just dumped in an uncontrolled or non-systematic manner. The other 75% is dealt with by controlled tipping. This means the waste is deposited, spread, and compacted into shallow layers, and covered with soil to assist decomposition and sealing. This method should ensure that loose litter does not blow about, there is no unpleasant odour, and flies and vermin do not breed to produce a health hazard. An alternative to tipping is the use of waste for land in-filling. Disused quarries, or land which is derelict by virtue of being low-lying and badly drained, or derelict as a result of spoil tips, can be reclaimed by refuse in-filling.

The shortage of suitable land for tipping purposes has caused some Local Authorities to consider alternative methods of waste disposal. Controlled tipping is usually cheap in respect of capital and labour costs, but it can become costly if highly priced land has to be purchased for future tipping. A way of avoiding this, and assisting land conservation, is to reduce the bulk quantity of the waste before tipping. This allows existing tips to be used for a longer time, reduces the need for new ones, and reduces labour costs for tipping operations. Waste can be pretreated by pulverizing, or mechanically breaking it down into smaller particle sizes, which can reduce the bulk by up to 33% by volume. Whilst the cost of a pulverizing plant is high, the salvaging of materials for recycling, and less waste to deposit can help to off-set the initial capital cost. Another pretreatment waste technique is incineration, which involves combustion in a furnace at a temperature between 950 and 1100°C to minimize corrosion and the emission of odours. This reduces the waste bulk considerably, and the process can reduce the volume up to 90% and the weight up to 60%, compared to untreated waste. There are environmental advantages to pretreatment. Much less residue has to be disposed of, and it is free from bacteria and wet organic matter that can cause putrefaction odours and gases. Also it is possible to use the heat energy produced for augmenting electrical generation or district heating, and this is already being carried out in some countries. It has been estimated that if all the household and trade refuse collected in the UK in 1974 had been incinerated and the heat used, this could have saved energy equivalent to 6 M tonnes of coal.

Some Local Authorities dispose of their waste by composting, but this is not carried out to a large extent (see Table 9.4). Basically this is an aerobic biological process whereby micro-organisms convert organic matter into a more stable material. The biological activity produces sufficient heat to eliminate any pathological organisms in the waste within a 4 day period. Waste pretreatment is necessary to remove materials such as metals, rubber, plastics, and glass, because they are largely unaffected by the process. Composting is usually carried out in some type of mechanized plant where each batch of waste remains for 5 days. One advantage of the process is that the waste volume is reduced by 50%. Separated materials can be sold for recycling or tipped, and the resulting compost can be used as a soil conditioner with some nutrient value.

Disadvantages of composting are the capital cost of the plant and the reluctance of farmers and horticulturalists to make use of the end product. This is probably because it is deficient in potash and inorganic nitrogen compared to farmyard manure, and may contain small amounts of toxic substances. If used on land, it requires high rates of application, possibly augmented by chemical fertilizers.

In recent times much more so-called bulky waste is being produced by householders. This includes obsolescent or worn-out furniture, bedding, re-frigerators, washing machines, and cars. There is also an increasing output of waste resulting from do-it-yourself or DIY activities involving discarding all sorts of materials, alterations to premises, and so on. Under the provisions of the Civic Amenities Act 1967, Local Authorites are required to provide sites where residents may dump this type of refuse free of charge. The intention of the Act is good, but it does not operate well because waste has to be conveyed by the discarder to the disposal site provided. This requires time, effort, some expense, and a sense of responsible behaviour from the public. Consequently, there is an increasing tendency on the part of some of the public towards the stealthy irresponsible dumping of all kinds of bulky waste on derelict land or in rural areas. The leaving of paper, plastics, bottles, and cans, as general litter at picnic sites, in parks and other open spaces is another form of pollutive public behaviour. This uncontrolled waste deposition should not occur, because it is a legal offence to abandon vehicles, or rubbish, or litter on open land or highways. Local Authorities are also empowered under the 1967 Act to clear up waste dumps, but many of them fail to carry this out on the grounds that they cannot afford the cost and labour involved.

Methods Used by Industry

An approximate estimate of the annual amount of industrial waste produced in 1973 was 110 M tonnes or 67% from the mining and quarrying industries, 12 M tonnes or 7.3% from CEGB power stations, and 23 M tonnes or 14% from other industrial sectors (Table 9.1). There are no published figures for radioactive wastes, but the amount is relatively small. The quantity and type of waste varies from industry to industry, but the major part consists of solid material, liquid slurries and effluent containing a wide range of suspended and dissolved chemical substances. The large amount of solid waste produced by the mining and extractive industries is disposed of by tipping on land or into the sea. An analysis of the methods in use is shown in Table 9.5. Although Table 9.5 does not include the stone and iron ore quarrying industries, it does show that about 75 M tonnes of waste per year is disposed of by tipping on land. Other wastes such as furnace clinker, blast-furnace slag, and copper, tin and zinc–lead slags have been omitted because they are mainly re-used in subsequent production. About 70% of this waste is not treated in any way and is tipped on land, or is used for land in-full and reclamation. The problems associated with this are discussed in Chapter 10 on land dereliction. It should be noted that about 26% of the above wastes are not

Table 9.5. Disposal of some solid industrial wastes (data from Department of Environment Building Reasearch Establishment 1974)

Type	M tonnes per year	Method	M tonnes Re-use	M tonnes Waste
Colliery spoil	45	Land tipping	8	37
Iron and steel making slag	13	"	11	2
Domestic glass refuse	2	"	some	2
Slate quarry waste	1.2	"	—	1.2
Refuse incinerator ash	0.6	"	—	0.6
Pulverized fuel ash	9.9	"	6.3	3.6
Colliery spoil	5	At sea	—	5
Calcium sulphate	2.1	Mainly at sea	0.006	2.094
China clay waste	22	In lagoons	1	21
Fluorspar mine tailings	0.23	"	some	0.23
Alumina waste	0.1	"	"	0.1
Overall total	101.13		26.306	74.824

tipped, but are used to assist in the production of materials such as bricks, concrete blocks, cement, and road and concrete aggregate.

Manufacturing industries produce wastes which are solid, semi-solid, liquid, or gaseous, and each category may contain toxic or non-toxic, flammable, and non-combustible constituents. There are no overall data available across all industries to show the quantities of waste or the methods of disposal. Some limited surveys have been carried out, and these at least provide some detailed information. The Local Government Operational Research Unit (LGORU) conducted a survey of the industrial wastes from 600 firms in the heavily industrialized area of Manchester and Salford in 1970. It was found that one million tonnes of waste per annum was produced, consisting of non-combustible sludge, dust, ash, brick, slag and excavated materials; and combustible paper, rubber, plastics, timber, sawdust, textiles and chemical materials. The quantities and methods of disposal are shown in Table 9.6. In this survey, 72% of waste was tipped on land, 16% of the intractable and dangerous waste was dumped at sea in sealed containers, and 8.6% was buried. Only about 3% of the waste was disposed of by the alternative methods of incineration or discharge into sewers. The

Table 9.6. LGORU Manchester/Salford Waste Survey 1970

Type of waste	M tonnes	Disposal method	M tonnes
Non-combustible	0.63	Tipped	0.720
Combustible	0.135	Dumped at sea	0.163
Intractable/dangerous	0.175	Buried on land	0.086
Flammable	0.015	Into sewers	0.016
		Incinerated	0.015
Total	1.000	Total	1.000

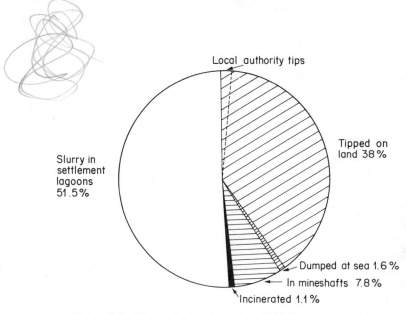

Figure 9.3. Disposal of toxic wastes (TCDWT Survey, 1970).

previously mentioned TCDTW survey also analysed the disposal methods used by the 1186 firms (Figure 9.3). The figure shows that 38% of waste is directly tipped on land under the control of three agencies. Local Authority tips received only 3.5%, waste contractors 7.7%, and 26.9% was deposited on industrially owned tips. In addition, sludges from lagoons, and waste put down disused mineshafts was also tipped on land. Therefore in this survey tipping was the method used to dispose of about 88% of the waste. The two relatively safe disposal methods, dumping at sea, and incineration, only accounted for 2.7% of the waste. The TCDTW report also provides details of the disposal methods of different types of waste, which are given in Table 9.7. It is reasonable to assume that wastes recorded under the headings 'contractors' and 'other methods' were dumped on land tips. The most toxic wastes would be acids and alkalis and indisputably

Table 9.7.   Disposal of solid and semi-solid wastes

Type of waste	Method of disposal							
	1	2	3	4	5	6	7	8
General factory	31.4	24.7	—	74.6	—	30.5	—	7.1
Inert process	62.5	58.9	99.7	5.7	96.7	53.3	98.0	91.0
Flammable	1.5	1.9	0.3	15.6	—	5.7	—	—
Acid and alkali	4.4	9.0	—	3.3	—	5.5	2.0	1.1
Indisputably toxic	0.26	5.5	—	0.8	0.3	5.0	—	0.7
Total % of all	3.5	24.5	7.8	1.1	1.6	7.7	51.5	2.4

Vertical columns show percentage of each type of waste.
Column 1, local authority tips; 2, other tips; 3, down mineshafts; 4, incineration; 5, dumped at sea; 6, contractors' tips; 7, slurry lagoons; 8, other methods.

toxic. The data shows that for acid and alkali waste 72.7% was dumped on land, 26.4% was in the form of slurry, and only 0.9% was incinerated. For toxic waste 96.5% was dumped on land, 3% at sea and 0.5% was incinerated. The LGORU survey dealt with general industrial waste, and the TCDTW survey dealt with industrial waste with some toxic content. Comparing the two surveys it seems that the disposal methods and the quantities involved are roughly similar. In both surveys more than three-quarters of the waste was tipped on land, and the smallest quantities were incinerated or dumped at sea.

The TCDWT report showed that about 60% of wastes were in the form of semi-liquid slurries, which were disposed of by dumping down mineshafts, at sea, discharged into sewers, or stored in lagoons. Slurries often consist of a complex mixture of dissolved and suspended chemicals in a varying volume of water and, or organic solvents. These slurries are often treated before disposal, for various reasons. They may have precipitating agents added to increase the rate of sludge settlement, or useful chemicals are reclaimed, or the acid content is neutralized. Often the slurry volume is decreased, which reduces the cost and the quantity of waste to be disposed. Certain very toxic substances can be treated to make their disposal safer, for example cyanide can be decomposed by chlorine, metallic salts can be precipitated as hydroxides of low solubility, and phenol can be biologically treated. Treatment of waste slurry usually helps to reduce the risk of environmental pollution. After treatment the sludge has to be disposed of, and most of it is tipped or dumped. The two surveys showed that between a quarter and a half of the wastes were dealt with by waste disposal contractors. In some cases three firms are involved, namely the waste producer, a transport contractor, and a disposal contractor. Close cooperation and communication between all the personnel involved is essential to reduce pollution risks. There can be serious problems involved in the disposal of very mixed types of chemical wastes, as illustrated by an incident at a toxic waste tip at Pitsea, Essex in 1975. A load of waste containing sodium hydrogen sulphite, sodium sulphide, sodium thiosulphate, sodium hydroxide and various organic compounds was dumped on a specific day. Fifteen minutes later, a second load was added containing aluminium oxide in sulphuric acid, and traces of lead, nickel, copper and titanium. When the two loads mixed, the chemical constituents reacted and sufficient hydrogen sulphide was released to fatally asphyxiate the lorry driver. The accident emphasizes the necessity for the disposal contractor and personnel involved to know the chemical composition of wastes, and to exercise great care in dumping them. At present a potential danger from tipped wastes exists, but few actual cases of pollution have been recorded.

Radioactive wastes produced from nuclear reactors and research establishments present special environmental disposal problems. The wastes consist of spent fuel rods and casings, fuel processing wastes, and various contaminated materials is solid and liquid form. They contain a range of about thirty nuclides such as Cs-137 and Sr-90, all of which have different levels of radioactivity. Therefore radioactive wastes must be processed and disposed of in such a way that no material is ingested or inhaled, and workers and the general public are not subjected to external ionizing radiation. Waste consisting of a mixture of nuclides is at present disposed of in different ways, depending upon the radioactive level or half-life of the nuclides. Liquid wastes of low radioactivity are discharged into the sea. Medium to low activity wastes may be burnt, or enclosed in sealed stainless steel containers, and buried on land or dumped into the sea. Since 1949, it is estimated that 64 000 tonnes of solid low activity wastes have been dumped into the sea from the UK alone. Highly radioactive wastes have to be stored on site in steel and concrete protective silos to allow the heat and radioactivity to decrease. The ultimate disposal of wastes containing nuclides like plutonium with very long half-lives is at present under investigation. Possible methods of disposal are the dumping and sealing of wastes below about 300 metres in granite, or some other type of impervious rock strata. Alternatively, vitrification or the sealing of wastes in leak-proof glass blocks, or dumping them in clay deposits or in salt mines may be used. Other views are that the wastes, suitably shielded and protected, should remain on the land surface where they can be continuously monitored for radioactivity. It has been estimated that by the end of this century, the nuclear waste silos in the UK will occupy a land area equivalent to two football pitches. In 1979 it was stated that Windscale accepts about 2000 tonnes of radioactive wastes per year. Present storage methods are not completely satisfactory, and there have been minor leakages from fuel storage silos at Windscale in 1976 and 1979. The safe disposal of waste radionuclides, which are being produced in increasing quantities, presents one of the most important environmental problems of today. It has serious implications for both the present population and future generations. It is also one of the reasons for the public concern and opposition towards the use of nuclear energy that is manifest in many Developed countries at present.

9.3 Land Tipping and Pollution

The total quantity of tipped waste material, and the area of land used for this purpose in the UK, is not known accurately. The amount is certainly several thousand million tonnes, and it is continuing to increase annually. Estimates for specific industries are the only guide available. The mining and extractive industries produce waste proportional to the amount of usable material removed. For example for every tonne of china clay extracted there are 6 to 8 tonnes of waste, and every three tonnes of marketable coal from a good seam produces at least one tonne of waste spoil. During the period 1963 to 1973 the amount of china clay spoil doubled, and other quarrying activities produced 75% more waste. Over the whole extractive industries excluding coal mining, the amount of mined material since 1955 has risen by 70%, to reach 329 M tonnes in 1968–69. Although the output of deep mined coal has been declining the amount of waste spoil has remained about the same. This is because the modern highly mechanized methods of coal production produce more waste than the former less mechanized methods. The National Coal Board (NCB) have forward plans to increase deep mined coal output to about 150 M tonnes per year in the 1980s, to open up new collieries, and to increase the extent of open-cast coal mining. All these indications show that mining and quarrying waste production will tend to increase in the future.

Solid waste tips can be a source of chemical pollution. If they contain organic refuse this is acted upon by naturally occurring bacteria, and gases such as ammonia, carbon monoxide, methane, carbon dioxide and hydrogen sulphide are produced. Carbon dioxide will dissolve in water to produce carbonic acid, so producing acidic conditions. Hydrogen sulphide has an objectional odour, and together with carbon monoxide is potentially poisonous, especially if concentrations of gas accumulate in the tip and are suddenly released. Toxic concentrations of gas should not occur if the tip is properly constructed and consolidated, with a surface sealing of soil. It is estimated that about 25% of tips in the UK are of the uncontrolled type where waste is carelessly tipped, and these may result in land pollution. Spoil heaps and land tips can also cause water pollution. The way in which this can occur is shown in Figure 9.4. The diagram shows a typical land-fill or buried tip used for the disposal of mixed industrial wastes. The tip is assumed to be carelessly sited in permeable substrate near extractable water supplies. The waste in the tip becomes effectively part of the natural environment, and subject to rain and soil water movements within the hydrological cycle. Rain water percolates through the waste, and removes some soluble pollutants and suspended particulate matter by the normal leaching process. The leachate solution and suspended solids then percolate downward and laterally into the soil around and beneath the tip. The contaminated leachate eventually reaches the water table, mixes with the ground water, and can enter an aquifer if it is present. Subsequent movement of the ground water carries the pollutants into the bore hole and stream, so contaminating water supplies.

This description relates to those site conditions that could cause maximum

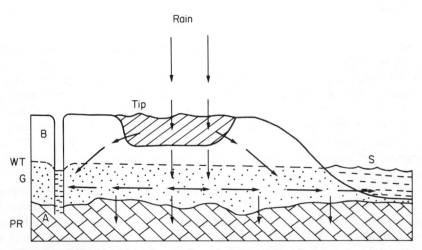

Figure 9.4. Pollution from a landfill tip: A, aquifer; B, Bore hole; G, ground water; PR, permeable rock; S, stream; WT, water table.

pollution of water supplies. To prevent pollution, the tip location should exclude the possibility of any leachate and toxic materials reaching the ground water. Preferably the tip should be saucer shaped and located in clay or shale, which is not readily permeable to water. If limestone is present in the geological strata this will also help to reduce the acidity of the leachate. The soil contains some natural barriers to the spread of any pollution. Aggregates of clay particles will strongly adsorb a variety of chemical ions from solution, and also act as a filter for suspended solids. There may be some percolation through the tipped waste, but it is possible to prevent pollution spread via the ground water if there is a basal sump or a drainage system below the tip. The drained leachate can then be drawn off and treated on site and piped away to sewers. When the tip becomes inactive it should be covered with a layer of impervious material to prevent the entry of rain. It is very important that the siting of solid and liquid waste tips should only be permitted where the type of geological strata allows only the minimum amount of leachate drainage away from the site. Neither should tips be located near underground or surface water supplies. Prior to 1972 there was irresponsible and indiscriminate tipping of toxic wastes, which caused periodic occurrences of water pollution. The Deposit of Poisonous Wastes Act 1972 and the notification that it requires has helped to control random tipping of toxic wastes. It has tended to lead to the establishment of large 'safe' tips that are efficiently operated and controlled by waste disposal contractors. One of the largest sites in the UK is located at Pitsea in Essex, and it extends over 500 ha. During the period June to August 1974, Pitsea received 85 013 m^3 of liquid wastes and 133 457 tonnes of other wastes, and these were transported from all over England. This prevailing practice where 30% of all toxic wastes are transported over 50 km for disposal, and 13% are carried over 150 km, is to be questioned. Toxic wastes or toxic substances of any type constitute a potential environmental hazard, and in the event of an accident, land and perhaps water pollution can result.

Some toxic wastes have been dumped in drums, when it has been assumed that if the containers are sealed and leak-proof there is no possibility of pollution. But metal drums have only a limited life, and when they become corroded the toxic contents can escape and cause land and water pollution. This can occur years after a dump has been closed, tested and pronounced safe. The use of drums for toxic wastes is now fortunately declining, and is being replaced by other methods, one of which is called encapsulation. This technique is suitable for both solid and liquid wastes. They are treated with polymers in a reaction vessel, and a lattice is formed that bonds with free chemical ions and encloses other insoluble waste components. The resulting thick slurry is run into landfill pits, and within three days it hardens into a polymer form that is impervious to water. This method should be used wherever possible for the disposal of all toxic wastes to safeguard possible pollution.

9.4 Waste Treatment to Reduce Pollution

A reduction in land pollution can be achieved by reducing the quantity of waste that needs to be tipped or dumped in the environment. There are three ways of doing this. The waste can be reduced in mass and volume by treatments such as incineration or pyrolysis. Less waste needs to be tipped on the land surface if more use is made of in-filling into quarries and mines. Thirdly, the amount of waste will also be reduced by the reclamation of re-usable materials, which reduces the need to use new finite natural resources.

Refuse incineration is being used mainly by those Local Authorities that do not deal with much industrial waste. This is because most municipal incinerators are not suitable for the combustion of toxic wastes. A few industrial incinerators are in use, but usually they deal only with free flowing liquids, and not solids and slurries that are produced in large quantities. Much more industrial use should be made of incineration, which is particularly suitable for disposing of toxic organic compounds, flammables, non-recyclable oils, and pesticides. One criticism of incineration is that it can produce atmospheric pollution. This can be minimized by careful control of the operating conditions and treatment of the furnace exhaust gases. The chemical composition of the waste should determine the operating temperature of the process within a range of 900–1400°C, in order to minimize toxic gas production. The gas stream should pass through cyclone cleaners to remove solids, scrubbers to dissolve out acid vapours, and electrostatic precipitators to remove any remaining small particulates. Incineration undoubtedly reduces the amount of toxic solid residue that needs to be land tipped. A new process called pyrolysis is in use in the US, and it produces a usable end-product and even less residual solid waste. The process consists of the destructive distillation of wastes in the absence of air. One variant produces a mixture of hydrogen, methane, carbon dioxide, carbon monoxide and ethylene, some of which can be used instead of natural gas as a fuel. Pyrolysis carried out at 900°C will yield twice the amount of energy put into the process for each tonne of waste. Tar, pitch, light oils, ammonium sulphate and carbon char are also produced. These are re-usable materials, for example carbon char can be used as a charcoal substitute. Pyrolysis is a more efficient process than incineration, and as soon as possible it should be developed beyond the pilot stage in the UK.

The reclamation of re-usable materials from raw refuse can be carried out before tipping, or prior to incineration or pyrolysis. Reclaimed materials that can be recycled are scrap ferrous and non-ferrous metals, paper and board, glass, tyres, textiles, used oil, discarded furniture and bulky equipment, and cars. The Department of the Environment Refuse Disposal Report 1971 states that only 1.1% or 220 000 tonnes of refuse was reclaimed by Local Authorities in 1966–67. The Authorities claim that there are so many problems involved that it is uneconomic to reclaim from domestic and trade refuse. The difficulties that they cite are unsuitable collection vehicles in use, longer collection times, higher labour costs, and the reluctance of householders to separate salvagable materials. Another factor that does not encourage recycling is the fluctuating capacity and market prices of the reclamation industry. An alternative to waste dumping and incineration was proposed by Dr A. Porteous in 1967. He called it cellulose hydrolysis, or the chemical decomposition of cellulose by interaction with water. Domestic and trade refuse contains cellulose in the form of paper, board and vegetable waste. Chemical hydrolysis of cellulose can be achieved by using sulphuric acid as a catalyst, to produce sugars. This is followed by a biological process, where the sugars are fermented by yeast to yield alcohol, ethanol, and carbon dioxide. The original refuse would be reduced in volume by 70% and in weight by 57%. One tonne could produce 0.115 tonne of ethanol with a market price of £11.5 at 1974 prices. By varying the fermentation process, the hydrolysed sugars could be converted into other alcohols, acetone, organic acids, polyols, and even single cell protein. This example shows that there are various possibilities for the recycling of refuse, if they can only be investigated.

Industry does not reclaim sufficient chemical waste. The UK reclamation industry is mainly concerned with only certain metals, oils, solvents, and acids. The extent of industrial reclamation is again constrained by economic cost. It is significant that many chemical firms are willing to pay for the long distance transportation of toxic wastes by disposal contractors, rather than treatment and reclamation on their own sites. They are apparently content to off-load mixed industrial wastes to contractors or the Local Authority, if either will accept them. It is obviously easier and less costly to mix all process wastes together rather than keep them separated. This practice often makes reclamation very costly, if not impossible, complicates the disposal problem, and increases the possibility of pollution. However there are indications that this attitude is changing because of several factors. The implementation of legislation since 1972, and the more stringent requirements of Local and Water Authorities is causing industrial companies to review their policies of waste disposal, and develop more on-site effluent treatment plants. The inflationary rises in the cost of raw materials, and increased effluent disposal and waste transport charges are causing manufacturers to consider reclamation and recycling more realistically. The Department of the Environment Waste Management Advisory Council is helping by co-ordinating the problems of waste disposal, recycling, and waste reduction. The DI Warren Spring laboratory at Stevenage also operates a UK Waste Materials Exchange Service for industrial and LA subscribers. This service provides regular information about waste suitable for recycling, so that the discarded waste of one firm can be purchased and re-used by another firm.

Various waste disposal methods have been discussed, and it is interesting to compare relative costs. Table 9.8 gives data taken from the Open University

Table 9.8.   Costs of waste disposal methods, 1975

Method	£ Disposal costs per tonne of waste
Controlled tipping	2–3
Pulverization	3–5
Pyrolysis	7.20
Incineration	9.10
Composting	9.44
Hydrolysis	£0.83 profit per tonne of waste with a 60% paper content

Unit 9-'*Municipal Refuse Disposal*', *1975*. It is assumed that plants operate for 5 days per week, and costs are based upon the capital plant outlay, labour and running costs, offset by any revenue from recycled materials. Tipping is clearly the cheapest method, but it is the most polluting. The methods that are more efficient and much less polluting are between 2.5 to 3 times more costly. If a determined effort is to be made to reduce land pollution, and consequent water pollution, then disposal methods must change. Waste disposers must become less cost conscious, and adopt a more responsible attitude towards the environment.

CHAPTER 10

Derelict Land

There appears to be no overall agreed estimate of the amount of land in Great Britain that is variously described as derelict. Surveys have been made, but they have not been based upon a common definition. The Department of the Environment (DOE) official definition is 'land so damaged by industrial and other development that it is incapable of beneficial use without treatment'. This limited definition includes disused spoil heaps, worked-out mineral excavations, land damaged by subsidence, and neglected and unsightly land. However, it excludes land in active use for any purpose, and land that is subject to planning permission for any future development. A better definition was provided by the Nature Conservancy Conference on 'Countryside in 1970' held in 1965. This stated: 'derelict land means virtually any land which is ugly or unattractive in appearance, e.g. spoil heaps, scrap and rubbish dumps, excavations, dilapidated buildings, subsided or war-damaged land; or any land which is neglected, unused or even under-used'. An alternative to providing a satisfactory definition is to describe derelict land in terms of the different types that exist in the environment.

Accepting the term derelict land in its widest sense, then it always presents an ugly and unattractive appearance. Land dereliction is caused by man's despoilation of the land surface, which can be brought about in five main ways.

1. Land may be used by an industry such as mining, quarrying or open-cast mining, with the result that excavations and unnatural spoil heaps are created. If these remain a permanent part of the landscape then the land becomes derelict.

2. Other industries and Local Authorities use land for tipping solid refuse, discarded bulky waste items, vehicles, toxic wastes and so on. Subsequently the tips are no longer used, and the sites become abandoned.

3. Land that was once in active use for various purposes, eventually becomes disused, and no effort is made to remove the buildings and equipment. Examples are abandoned industrial sites, closed railway and canal routes, farm buildings, mines, and former Ministry of Defence sites such as aerodromes, camps, stores depots, gun emplacements, and block houses.

4. Houses, shops, and other premises in towns and cities that are no longer inhabited or used. This type is often called urban dereliction, or inner city dereliction.

5. Rural land that is not under cultivation and presents an unattractive appearance. This category includes land that has become unusable through

Table 10.1.   Derelict land area in GB—MHLG 1966–68

	Area (ha)	%	Justifying treatment	'Hard core' derelict	%
England (1968)	37 500	73	23 000	14 500	76
Wales (1968)	8 005	15.5	5 370	2 635	14
Scotland (1966)	6 070	11.5	4 045	2 025	10
Total	51 575		32 415	19 160	
			(63%)	(37%)	

mining subsidence causing sinking and poor drainage. Also uncultivated land that becomes used for the indiscriminate dumping of litter, building waste, cars, worn-out domestic equipment and furniture by the general public. In all these types of derelict land the condition and appearance either remains unchanged, or it deteriorates over varying periods of time due to neglect.

The use of statistics from various sources to establish the amount of derelict land in GB reveals that the compilers used different bases and definitions. Government figures are based upon the limited official definition previously described. A Ministry of Housing and Local Government (MHLG) report entitled 'New Life for Dead Lands', 1963, estimated 60 700 ha of derelict land in England and Wales. The Civic Trust 'Derelict Land' study in 1964 was concerned with 'derelict areas which are likely to remain offensive for a long time unless we are prepared to deal with them now, simply because they are offensive', and it classified the land into three types. Derelict land in urban areas was estimated at 14 570 ha, industrial land in less populous areas at 46 140 ha, and the net annual additional land due to mineral workings and tipping was 1416 ha. This is a total of 62 126 ha for England and Wales. Derelict Land Units (DLUs) for Wales and Scotland have estimated their areas as 7690 ha in 1969 for Wales, and 14 970 ha for Scotland in 1972. If the Civic Trust total of 62 126 ha for England and Wales is accepted, the addition of 14 970 ha for Scotland gives an overall total of about 77 000 ha of derelict land for GB. The Nature Conservation Conference in 1970, using its own definition, estimated the land area at 101 775 ha. MHLG statistics of their official derelict land areas are shown in Table 10.1. An approximate estimation of derelict land in GB is therefore probably between 60 000 and 100 000 ha, based upon the varying figures available. The real figure could be much greater, especially if the wider definition of derelict land is used.

The total amount of derelict land does not remain static, but increases annually. The Civic Trust study states that about 4850 ha per annum are taken up for the working of minerals, and the tipping of waste spoil. Some indication of the extent of this is shown in Table 10.2. The table shows that only 70% of the land used by the extraction industries is eventually reclaimed. The remaining 30% remains permanently derelict. Further additions of derelict land result from the tipping of general refuse and waste by both industry and LAs, the tipping of other industrial solid and liquid waste, and indiscriminate dumping by the public. No quantitative figures appear to be available for these additional increases.

Table 10.2. Annual land usage for coal and mineral extraction, 1958

Type of operation	Land used (ha)	% restored	Left derelict (ha)
Open-cast coal mining	2710	95	135
Open-cast iron mining	240	95	12
Sand and gravel working	810	50	405
Clay and shale working	180	15	150
Other surface working	405	2	397
Deep mined coal spoil	305	—	305
Other tips	200	75	50
Total	4850		1454 (30%)

10.1 Permanently Derelict Land

Environmentally, it is important to consider land that seems to be accepted as permanently derelict. Most of this derelict land is a legacy from the past, and it mainly resulted from the activities of the extractive industries. It is spread throughout England, Wales and southern Scotland, and can be correlated with the industrial distribution shown in Figure 10.1. Particularly bad derelict areas occur in Cornwall, Lancashire, Durham, Staffordshire, South Yorkshire and South Wales, and these areas account for over 60% of the total derelict land in England. This dereliction was produced largely as a consequence of the industrial revolution of the late eighteenth and nineteenth centuries. This caused increased demand for coal as a fuel, iron and other minerals for industrial and railway development, gravel for roads and railways, and brick clay for building. In the nineteenth and early twentieth centuries there was no planning and control of the use of land, and mineral extractors had no obligation to restore or reclaim land during or after their operations. Consequently through the wanton misuse of land, thousands of hectares gradually became derelict, and this contributed to the general increase of environmental pollution in this period. Industrialists, the Government, and regrettably the population accepted dereliction as inevitable, and it became part of local life. It went on increasing and little or no attempt was made to recognize, let alone contain or decrease, land dereliction. The natural landscape became completely changed by conical spoil heaps and excavations, and more and more argicultural land was removed from food production. One third of the derelict land in England, and over one half in Scotland and Wales contains spoil heaps. One heap may consist of over 5 M tonnes of waste, and rise up to 90 metres high. Spoil heaps not only destroy the natural amenity quality of the locality, but they are a source of air and water pollution, and may be physically unstable. This was graphically illustrated in October 1966 at Aberfan in South Wales. Slag and mud slurry from one of three nearby colliery tips engulfed the village and resulted in 144 deaths, of which 116 were children in the local school. Abandoned quarries, mineshafts, and hillside excavations also affect the natural environment, and these account for 12 140 ha or 27% of the area of derelict land in England and Wales. If located below the soil water table, they fill with water and can contain putrefying material as well as constituting a public safety hazard to both adults and children.

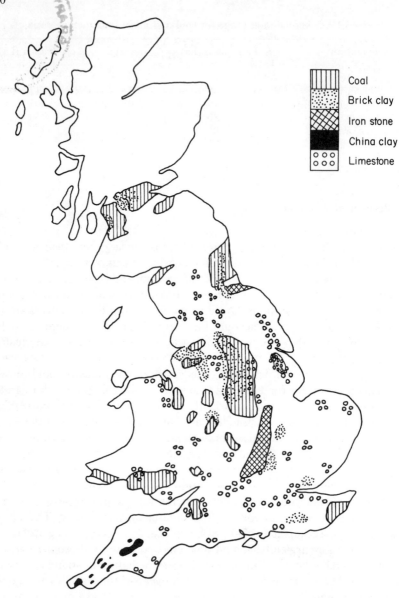

Figure 10.1. Coal and mineral working areas in GB.

Apart from the pollution effects discussed, derelict land areas often have social ill-effects that lead to their decline. It is bad enough that people have to live in squalid, ugly surroundings but it is worse if local employment declines. For example, there can be a reduced demand for coal and mineral products in periods of economic recession, workings become abandoned as resources are worked out, or they become uneconomic to operate. The local population's standard of living declines, and even if they have employment they lose their respect and appreciation for the environment, and become apathetic. So there is a general lowering of living standards causing untidiness, unkempt houses and gardens, and indiscriminate dumping of rubbish. The combined effects of dereliction alter the whole environment and its inhabitants, and they are self-perpetuating. In the 1920s and 1930s many derelict areas became known as depressed areas, which aptly described their visual, social and economic degradation. Subsequently, governments attempted to encourage new industrial development in these areas and so reduce the unemployment rate. But little attempt was made to improve the dereliction and aesthetic appearance of the areas, so consequently industrialists and their employees often rejected the depressed areas as sites for new development.

In deep coal mining areas, sinking of the land surface, or subsidence, takes place to a varying degree. It is caused by the removal of the coal seams, and is especially severe where no precautions are taken, and strata faults are present at the edges of the worked area. Subsidence causes damage and destruction to buildings, railways, roads, and sewers, and disrupts the land drainage, causing the area to become marshy and subject to periodic flooding or flashes. Farming land may become unused and derelict, and the general appearance of the neighbourhood declines. Some subsidence is always present in a mining area, but under the Coal Mining (Subsidence) Act 1957, the National Coal Board (NCB) must accept liability for this and pay compensation.

On a lesser scale, the Ministry of Defence and nationalized transport are responsible for dereliction. Large tracts of rural countryside have remained requisitioned by the War Department and the Air Ministry since the last war. It is claimed that out of 242 820 ha held in 1970, 141 650 ha are required for training purposes, and the rest for a possible future emergency. The land is often not properly maintained, it is not productive, and may contain obsolete and abandoned buildings, runways, etc. The British Railways Board has progressively reduced its track mileage since 1963, as a result of the Beeching policy to close uneconomic routes. By 1971, 13 450 km of track had become disused as well as hundreds of stations, freight marshalling yards, and other buildings. About 54 430 ha of derelict land now consist of disused railway tracks alone. Similarly the British Waterways Board, since its creation in 1948, has reduced its length of open navigable waterways by 1240 km. Most of these waterways have become derelict, and a further 1600 km have not been improved but merely maintained, because of an inadequate budget of only a few million pounds per year. Unless Government policy changes it seems inevitable that further canal stretches must become neglected and derelict in the near future.

Derelict land and buildings can be found in large cities and towns that expanded rapidly during the last century. Urban dereliction is often the result of changes in technology or industrial developments, or the need to improve people's life-style. For example, factories and processes may become obsolete, production ceases, and the buildings become disused and derelict. Housing is adjudged to be sub-standard, and so people move out into new housing estates or flats, often located outside the town or city centre. Cities which are sea ports have been affected by a world reduction in the size of passenger and merchant shipping fleets, and the development of modern container cargo handling methods. This has caused dock facilities to become out of date, and eventually disused and derelict. Short-term dereliction exists where buildings and land have been evacuated, because the site has been earmarked for future redevelopment by Local Authorities or property developers. Alternatively some urban dereliction was caused by war damage, and in some towns and cities this has existed permanently since the 1940s. All these types of land pollution can be seen in central urban areas, and the derelict buildings present a desolate and unsightly appearance, and may constitute a health hazard. This is especially the case if the buildings are occupied by vagrants and squatters, whose precarious existence usually implies unhygienic and squalid living conditions. Where no action is taken to clear these areas there may be a tendency for the derelict land areas to spread. One of the causes of this has been the inflation of land values. Landowners have purposely refrained from any action and waited for the land value to increase over the years. The overall effect is that the total urban built-up environment increases in size, yet it has unused derelict land within it, and new building development takes place on the periphery, utilizing productive agricultural land. A determined effort should be made to promote the clearance of derelict buildings, and the open spaces created could be landscaped or used as temporary car parks pending redevelopment. The DOE has now become more conscious of inner city dereliction and is attempting to encourage reclamation and new development. This will not only reduce the scale of dereliction, reduce urban sprawl into rural areas, but also improve the quality of the environment for urban inhabitants.

10.2 Land Redemption

After the Second World War it became Government policy to promote a reduction in land pollution. Perhaps a significant change of attitude was shown by the fact that depressed areas became known as development areas. Land redemption can be carried out in three ways: land restoration where the land is returned to agricultural production; land reclamation, resulting in its use for private, industrial or public purposes; and rehabilitation where land is assimilated into the surrounding landscape by surface treatment.

The Town and Country Planning Act 1947 brought tipping and mineral working under Local Authority control, and required the ironstone extractors to carry out land restoration. The Mineral Working Act 1951 established the Ironstone Restoration Fund, whereby money is provided jointly by the industry and government to meet the cost of land restoration. The fund has worked successfully, and in Northamptonshire alone over 2000 ha of land has been restored since 1951. Similarly the NCB is required to restore land used for open-cast coal extraction. The NCB have successfully developed mining techniques as a continuous operation where top soil, sub-soil, and overburden are removed and tipped separately, the coal is extracted, and the overburden and soil are replaced in the correct sequence. This usually means that the land is out of agricultural production for an average of 3–4 years, and grass can be satisfactorily grown about 2 years after restoration. The NCB also replaces fences, and has undertaken mature tree transplanting as part of the restoration policy. However, other industries involved in china clay, brick clay, gypsum and salt extraction have no such statutory reclamation requirement, and so their activities continue to cause increasing land pollution. Local Authority planning permission for new extractive operations should always include after-treatment and land restoration. It appears that some Authorities are not stringent enough in their planning requirements, or in the monitoring of the extractive operations. Also, Local Authorities should be willing to change planning conditions in the light of altered environmental circumstances. This is often not carried out at present because of the fear that compensation may be claimed against them. There should be a national policy in the future, implemented by all planning authorities, to ensure that all open-cast mineral operations are followed by obligatory land restoration.

The pollution caused by mineral working tips has too often been regarded as an intractable problem. An example of this is shown by the china clay extraction industry centred mainly in Cornwall. The workings are steep open-sided pits up to 60 m deep, often covering an area up to 4.2 ha. The extraction of 1 tonne of china clay by powerful jets of water produces 6–8 tonnes of quartz, sand, and miscellaneous rock waste, which is deposited in white, conical shaped tips up to 45 m high. In the St. Austell locality, there is an estimated tip area of 1200 ha spread over 7.8 km^2 of countryside. These ugly tips dominate the landscape, and they tend to be wet, unstable, and badly drained. Little or no effort has been made to treat the tips, or to improve the topography of this so-called lunar landscape. Reclamation is not considered to be feasible, but rehabilitation could be carried out if the tip slopes were made less steep and the tops were flattened. Vegetation could then be planted and the tips would blend into the landscape. Alternatively the amount of heaped spoil could be reduced. Some local use of waste is already made for the production of concrete paving slabs, fencing posts, and building blocks. It has been estimated that up to 80% of the spoil could be utilized for building products, and used as a source of sand and gravel. There is a low demand for these products in Cornwall, but the waste could be transported to the south-east, where there is a continual shortage of such building materials. Deep mine colliery spoil heaps are another major dereliction problem, and they probably occupy about 8000 ha in England and Wales. The NCB has no legal obligation to deal with permanent tips that existed before coal nationalization took place in 1948. However as a result of the Aberfan tragedy, the NCB is attempting to tackle the more potentially dangerous tips. They are flattening the contours of old tips, and using modern mobile tipping methods instead of static overhead ropeways for new tips.

In a few areas of the country such as Lancashire, the former West Riding, Staffordshire and the North East, imaginative land reclamation schemes have been carried out since the 1950s. For example the Stoke-on-Trent Council transformed 52 ha of permanently derelict land in the late 1960s. They employed landscape architects and other consultants to transform a derelict area of colliery spoil heaps and pottery marl tips into an amenity and recreational area. A hilly wooded sector provided paths and trails for walkers, cyclists and horseriders, and other sectors were laid out as football and sports areas and a boating lake. The Council also planned to link this development to other reclaimed areas by a system of greenways to be landscaped out of disused railway tracks. In Lancashire, there are areas littered with colliery spoil heaps and derelict buildings. Here land affected by subsidence, chemical waste, and rubbish heaps has been reclaimed. New housing and industrial estates have been built as well as recreational areas for games and athletics. These two examples illustrate what can be achieved if Local Authorities are prepared to assume responsibility for reclaiming present derelict land. The DOE will provide assistance for approved schemes in the form of specific grants from central funds. These can be up to 85% of the cost in development areas, and a minimum of 50% in non-priority areas. In addition, Authorities may get more money through the annual rate support grant, which together with a reclamation grant, can amount to 95% of the total cost. Spoil heaps can be rehabilitated by drainage, levelling, and in-fill operations. Provided that substrate instability and surface erosion can be reduced, it is possible to grow up to 15 different tree species on the sloping sides of tips. They do not necessarily require top soil, and will grow on raw shale and heavy overburden. There are now well established afforested tips which merge very pleasantly into the surrounding natural landscape.

The tipping of refuse as land in-fill under controlled conditions can provide restored or reclaimed land. Both private companies and Local Authorities have reclaimed old refuse tips for use as playing fields, picnic areas, car parks, and housing areas. Subsidence will always occur to some extent in mining areas, but the ill-effects can be reduced by better underground mining techniques. There should be much greater use made of back-stowing methods, whereby spoil remains underground and is used to fill the space left after the coal removal. This can minimize subsidence and also reduce the area of surface spoil heaps. It is encouraging to note that in the development of the new Selby coalfield, the NCB has given assurances that there will be no spoil heaps. Waste will be used for landscaping and the raising of river banks, and every effort will be made to minimize subsidence and protect buildings.

Derelict quarries and other dry excavations have been used for the tipping of fly ash, household and other waste materials, and then reclaimed. The surface is covered with top soil, and then grassed over or afforested. Flooded or wet quarries have been filled in where the surrounding geological strata prevents any possibility of underground water pollution. A factor that often prevents quarry reclamation is a shortage of suitable in-fill material in the locality, so necessitating transport from other areas and additional cost. Another type of reclamation, depending on the type of quarry, is to produce a recreational facility for boating, sailing, fishing, and swimming, with suitable safeguards for public safety. An interesting reclamation project was started in 1966 in the Peterborough area, through the close co-operation of brick manufacturers, British Rail and the CEGB. The Central Electricity Generating Board power stations in the Trent valley had a major waste disposal problem, because they produce millions of tonnes of pulverized fly ash each year. The Peterborough brick companies are continually excavating clay, and they were willing to reclaim their pits which already had a capacity of 30 M tonnes. The power stations and the pits are 97 km apart, but British Rail agreed to run seven trains a day, 6 days a week to carry the ash to the clay pits. On site, the ash is mixed with water to form a slurry and this is pumped into the pits and solidifies. The reclaimed area is able to support housing or industrial buildings. When the project began it was estimated that it would take 20 years to fill the derelict pits that existed at that time, but of course new pits are being worked continuously. The Peterborough project illustrates how co-operation can achieve land reclamation and a reduction of tip dereliction. Not all land redemption is carried out by Local Authorities and industry. There are increasing numbers of voluntary associations and helpers such as the Civic Trust, Community Service Volunteers, and the National Trust, who are helping to reclaim and restore derelict land. Former British Rail tracks and buildings have been purchased by local societies of steam railway enthusiasts in different parts of the country. Their efforts have restored track and buildings to a working condition and provided local travel and amenity facilities instead of dereliction. Other disused tracks have been used for foot paths of nature trails. The Inland Waterways Association as a national voluntary body have removed rubbish and weeds from canals, repaired tow paths and locks, and restored stretches of waterways to active use for fishing and cruising. Numerous other small projects have cleared rubbish from disused land, streams, and ponds under the banner of various 'Clean-up' campaigns. Since 1976, unemployed people under Local Authority supervision have been used for projects to improve the local environment.

The total amount of land redemption that is being carried out at present is not accurately known. Some information is available from the DOE (see Table 10.3), but these figures relate only to the Ministry's official definition of derelict land, and that which is grant aided. The stated figures of land justifying treatment date back to 1969. If these are accepted as an indication of the relative position, then

Table 10.3. Derelict land reclaimed in GB, 1969–72

| Area | Land justifying treatment (ha) | Land reclaimed (ha) | | | | Time of reclaim all land (years) |
		Total	Average/year	Time (years)	
England	23 000	6600	1650	14	23
Wales	5 370	1480	370	14.5	22
Scotland	4 045	720	180	22.5	34
Total	32 415	8800	2200		

the present rate of land reclamation is very slow. If all the derelict land in GB were to be reclaimed at the above rate, then it would probably take 40 to 50 years at least. The cost of such a programme would be considerable. The average annual expenditure from Central Government funds for land reclamation was given as £ 12 M in 1972, which is a cost of £ 4400 per hectare. To reclaim all land justifying treatment could require an outlay of about £ 200 M at present day costs. This is about 18% of the revenue received from all vehicle licences in 1976.

Obviously much needs to be done to reduce national land pollution caused by dereliction. Localized action by many individuals and organisations is helping to improve the position, but a nationally coordinated approach has a low priority in government policies. The DOE has a Central Group for Derelict Land Reclamation briefed to 'keep under continuous review the progress of derelict land clearance in England and to identify obstacles in the way of progress, and to advise Ministers on how these could be removed'. There are also Derelict Land Units for Wales and Scotland. What is required is a more determined Government commitment to promote action at local level. County Planning Authorities are required to produce structure plans for their areas, and these should include positive proposals to deal with derelict land of all types. The plans require approval by the DOE so that the adequacy of derelict land redemption proposals could be monitored. Once approved, then the DOE should be prepared to grant aid the cost to Local Authorities, within an agreed time scale.

CHAPTER 11

Agricultural Land Pollution

11.1 Farming Practices

The productivity of the UK agricultural industry has increased considerably during the last 20 to 30 years. This has occurred in spite of a reduction of over 50% in the labour force, and a decrease of 7% in the area of land under cultivation. The increase in home grown food has been achieved by the application of modern science and technology to farming practices. Crop husbandry methods now involve the use of highly developed machinery and the application of inorganic fertilizers and organic pesticide chemicals. Animal husbandry methods have also changed, with increasing emphasis upon the indoor housing of cattle, pigs, and poultry within controlled environmental conditions. These fundamental changes in agricultural practice have caused new environmental problems to arise, especially in relation to organic waste disposal and the use of chemicals.

Fertilizers

Before the adoption of modern farming methods, the remains of crops and animal manures were returned to the soil. The dead organic matter was decomposed by soil micro-organisms and converted into humus. Humus is a complex of organic compounds and an essential constituent of all natural soils, because it helps water retention, soil water movement, provides food for soil micro-organisms and animals, and is a source of plant nutrients. Specific bacteria decompose organic molecules to form nitrates, phosphates, potassium (potash), and sulphates through natural biological cycles. The development of 4 year crop rotational practices in the nineteenth century was to assist the maintenance of soil fertility. However, in the twentieth century, modern farming techniques include more intensive cropping with little or no rotation, and the development of monoculture where cereals are grown continuously on the same land for several years. Treatment of the land with manure containing straw is now largely replaced by slurry. Therefore man has effectively changed the agricultural ecosystem, with the result that there is less humus in the soil, causing a deficiency of plant nutrients, and a change in the weed, pest, and disease balance. To redress the balance and to boost crop yields, the modern farmer relies heavily on chemical soil and crop treatment. Considerable applications of NPK (nitrogen, phosphate, potassium) inorganic fertilizers are used, which are quick and easy to apply and are in a chemical form that is readily available to crops, without microbiological action. NPK derived from humus is slowly released into the soil, and most of it is taken up by the crop. But where an inorganic NPK fertilizer is used on a soil with a low humus content there is usually an excess above crop requirements, and some fertilizer is leached out into water courses. This takes place particularly where a heavy fertilizer treatment is used on a heavy wet soil in seasons with a heavy rainfall. Nitrate is readily leached out of soil, and there are now many instances of this types of water pollution. For example in 1977 the River Thames contained five times as much nitrate as it did in 1948. Potassium is largely retained in the soil, and phosphate to a considerable extent, especially where the humus content is good. But a Centre for Agricultural Strategy Report in 1978 stated that over one third of the 377 000 tonnes of phosphate used annually is now leached out of the soil. Increasing nitrate and phosphate water pollution is a cause for environmental concern, as it causes excessive aquatic plant growth, and excess nitrate in drinking water may adversely affect human health.

Intensive Livestock Farming

The modern trend in livestock farming units is towards intensive methods of production. The well established practice of animals spending most of their lives outdoors or free-ranging has been shown to be inefficient, in respect of the conversion of food into energy for growth and weight increase. Animals are a production unit in a modern agricultural system. They are reared to produce meat, eggs, milk, butter, and cheese for human consumption, by the most efficient and economic methods. This commercial approach has led to the practice of keeping large numbers of animals in an indoor closed environment, where the temperature, feeding, and animal health conditions can be largely controlled. In the UK in 1977 it was estimated that 13.8 M cattle, 7.7 M pigs, and 134 M poultry spent all or some of their lives indoors. Intensive housing methods are claimed to produce animal fattening, or weight gain at maximum rates, so reducing the cost and time of rearing to a minimum. But these methods produce considerable quantities of organic waste that accumulates and has to be disposed of. It has been stated that the agricultural industry produces over 120 M tonnes of organic slurry per year, which is more than the amount of sewage produced by the entire UK population.

Farm waste is derived from three main sources. Manure comprises animal excreta, urine, and bedding material, but on modern farms there is less use of straw and similar bedding materials so the animal waste is usually produced in the form of slurry. Intensive cattle housing and milking requires the maintenance of a high standard of animal hygiene. This involves the regular hosing down of buildings, milking parlours, and farm yards, which produced quantities of very liquid waste. The third source of highly polluting organic waste is from silage production. The winter housing of cattle requires controlled feeding with highly nutritious feedstuffs, consisting of silage and processed concentrates. Silage is produced in the summer by bacterial action upon freshly cut grass, and is stored in clamps or silos. Silage production results in the formation of varying quantities of effluent, that has a very obnoxious and long-lasting smell, a high organic content, and an acid pH. The amount of effluent produced depends upon the method of silage production. For example Ministry of Agriculture, Fisheries and Food data shows that an increase of 5 to 10% in the dry matter content of grass produced up to a 50% decrease in the quantity of silage effluent produced. This can be achieved by allowing the cut grass to wilt before use, and this procedure also improves the nutritional quality of the silage. The use of tower silos rather than clamps for silage making also reduces the amount of effluent. Poorly made silage can produce up to 228 litres of effluent per tonne of silage, and in 1975 approximately 7.73 M m^3 of effluent was produced in England and Wales. There is no doubt that silage effluent is a serious agricultural pollutant, and the high organic content causes it to be one hundred times more polluting than other farm wastes. Numerous cases of water pollution occur where this effluent has been allowed to enter streams and rivers. Animal intensive housing methods therefore produce quantities of polluting organic waste, the disposal of which causes environmental problems.

At present most farm waste is not treated prior to disposal. This is particularly the case on small and medium-sized farms where the capital and labour costs of treatment would be high. The usual method of waste slurry disposal is to spray or mechanically spread it on the land at the appropriate season of the year. Where any slurry treatment is carried out, it usually consists of separating the liquid fraction from the organic solids. The liquid is pumped on to the land and the sludge is stored for land disposal later. These methods may appear to be satisfactory because they are returning or recycling nutrients and organic material back to the land. However, slurry can have a high water content, a high nitrogen content, and contain phosphates and organic pesticides that were present in the original animal food. Where this slurry is put on to heavy land or applied in wet seasons when the soil water content is high, it may be a source of pollution. This is because the volume cannot be readily absorbed and retained in the soil, so there is considerable surface run-off, and drainage of ground water containing pollutants into water courses. Pollution is further increased if slurry and silage effluent are discharged directly into streams and rivers. This should not take place, because farmyard effluent is classed as trade waste under the Control of Pollution Act 1974, and its discharge is subject to a consent from the Water Authority. To reduce pollution in the future it may be necessary to require large livestock units to treat effluent before disposal. The effect of this would be to reduce the water and organic content of the waste, so that there is less risk of pollutants reaching surface waters. A reduction of all forms of agricultural land pollution is needed in the future to safeguard food and water supplies. More emphasis needs to be given to regular soil testing for nutrients, and the use of precise fertilizer application rates for specific crops. This would lessen the possibility of soil nutrient residues accumulating, and also reduce the total national fertilizer demand, the production of which requires the use of finite natural resources and energy.

11.2 Pesticides

A pesticide is a general term used for any substance used to control fungi, insects, or other animal pest that attacks food sources useful to man. Pesticides can be classified into the three main classes of insecticides, fungicides, and herbicides (or weed-killers). There are also rodenticides used for controlling vertebrate pests, nematicides for eelworms, molluscicides for slugs and snails, and acaricide for mites. Pesticides have been used to a varying extent since about 1850, but prior to the 1940s they were used in relatively small amounts on a local scale, and they caused no detectable harmful effects. Often natural plant derivatives were used, such as nicotine from tobacco species, pyrethrum from tropical chrysanthemums. and rotenone from the root of the derris plant. These substances are still in use to a limited extent and are regarded as 'safe' pesticides. Other non-organic chemicals were also used, including Bordeaux mixture (containing copper (II) sulphate), and preparations containing mercury, lead, zinc, and sulphur. These pesticides produce stable residues that can persist or accumulate in the soil. They may be removed by leaching from fields and orchards, and enter streams and rivers causing death to algae, fish, and other aquatic life. However, these inorganic chemicals are not considered to be major pollutants of land and water.

From the 1940s onward, the development of numerous synthesized organic compounds completely changed the basis and operation of pesticide control. Prior to 1940, the non-organic compounds in use were contact or surface pesticides. These did not penetrate plant tissues to any extent, and because they were susceptible to weathering their effect was very variable. Many of the new organic compounds are described as systemic pesticides, which means that they penetrate into plant tissues and are much more effective in their control action. The active chemical ingredient inside the plant is able to kill fungi in the tissues, and as it is absorbed by insects feeding on the plant tissues and fluids it will effectively control pests. The first synthesized pesticides to be developed were mainly insecticides, and these were followed by herbicides, so that by the 1950s a range of systemic insecticides and herbicides were available. Later in the 1960s effective systemic fungicides were developed and came into common use.

Table 11.1 lists some of the numerous pesticide substances in use. Insecticides such as DDT, which came into use in the 1940s and early 1950s, were welcomed as the complete solution to pest control. They were widely used for the control of vector insects carrying sickness, typhus, and typhoid. Other insecticides were used in agriculture and horticulture, because the control they produced increased crop yields and quality. So an increasing use of pesticides became part of the newly developing agro-technology aimed at increased efficiency and productivity on farms and holdings.

During the 1960s it was realized that any one pesticide did not entirely eliminate all the species population of a target pest, and so the concept of a

Table 11.1. Some pesticides in common use

(a) Insecticides		
1 Natural organics	Nicotine	
	Rotenoids (Derris)	
	Pyrethrum	
2 Synthesized organics		
Organochlorine	DDT	Aldrin
	gamma-BHC (Lindane)	Dieldrin
	Chlordane	Endrin
Organophosphorus	TEPP	Dimethoate
	Metasystox	Menazon
	Malathion	Chloropyrifos
	Dichlorvos ('Vapona')	Phorate
Carbamates	Carbaryl ('Sevin')	Propoxur

(b) Fungicides		
3 Surface types	Phenyl/alkyl mercury	Captan
	Thiram	Dinocap ('Karathane')
	Zineb	Quintozene
	Maneb	Rovral
4 Systemic types	Benomyl	Triarimol
	Thiophante-methyl	'Dowco'
	Carboxin	'Milstem'

(c) Herbicides		
5 Non-selective	Simazine	Diquat
	Glyphosate	Paraquat
	Monuron	
6 Selective		
Growth regulators	MCPA or B*	Mecoprop
	2, 4-D*	Dichloroprop
	2, 4, 5-T*	
Growth retardants	Maleic hydrazide	Dalapon*
	Chloromequat	TCA

*These herbicides can be used non-selectively during the growing season
Quotation marks indicate Trade Mark names.

resistant species was recognized. The result of this was that pesticide control for a particular species became increasingly less effective over a varying period of time. This has produced a number of trends in the use and development of pesticides. Continuous research and development is required to produce new pesticides that will provide more effective control. Alternative products of differing efficacy are developed for the control of a similar group or type of pest, for example Aldrin and Miral for soil pests. New chemical formulations are produced to provide more effective control of a specific pest, for example Mecoprop for chickweed, which is unaffected by general herbicides such as 2, 4-D, or MCPA

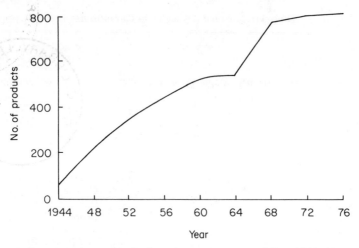

Figure 11.1. Agro-chemical products on approved lists 1944–1976
(from the Royal Commission on Environmental Pollution *7th
Report* (HMSO, 1979)).

Consequently the number of commercial products available has increased
dramatically over the last 20 years (see Figure 11.1).

The use of pesticides has been intensified in attempts to control resistant species
and produce more effective overall pest and disease control of crops. Repeated
applications of the same pesticide are made to compensate for pests that were
'missed last time'. Higher concentrations of chemicals are used than those
recommended, to try to kill resistant types. 'Insurance' or 'calendar' applications
are made at specific times of the year irrespective of whether there is any evidence
of pest and disease damage. The extent of pesticide use in Britain (see Table 11.2)
is shown by the fact that in 1977, 55% of farmers sprayed cereal crops once to
three times per season, and 42% used four to six applications. The corresponding

Table 11.2. Estimated annual average quantity of pesticide active ingredient used in 1971–75. (From the Royal Commission on Environmental
Pollution *7th Report*)

Type	Million Spray hectares	%	Tonnes per year
Insecticides			
Organochlorine	0.15	1.2	132
Organophosphorus	0.84	6.6	419
Others	0.12	0.9	779
Seed treatments	3.72	29	565
Fungicides	1.9	14.8	2 194
Herbicides	6.02	47	15 712
Other pesticides	0.05	0.4	1 960
Total	12.8		21 761

figures for other arable crops were 43% and 46% respectively. Comparable figures for world usage of pesticides are herbicides 43%, insecticides 32%, fungicides 19%, growth regulators 3%, and miscellaneous agro-chemicals 3%. World usage of total pesticides shows a very unequal distribution, with the US using 45%, Western Europe 25%, Japan 12%, and the rest of the world 18%.

As increasing amounts of pesticides were used during the 1950s and early 1960s, doubts were raised concerning the effects of these chemicals upon the environment. Rachael Carson was one of the first people to become aware of the dangers of pesticide pollution. In the now famous book '*Silent Spring*' in 1963, she produced evidence of toxic effects upon wild life and potential hazards to man, in their related ecological context. What had not been realized was that pests co-exist within biological systems with other forms of life. The elimination of pests must inevitably produce changes, and disrupt the balanced natural cycles and food nets within ecosystems.

Pollutive chemicals can enter plants in various ways. If plant foliage is sprayed with an organic insecticide as a control measure, some of the compound is absorbed into the leaves and may still be present when the crop is harvested for animal or human food. Alternatively up to 50% of the spray can be washed off the leaves by rain, and so enter the soil. Other chemicals may be sprayed directly on to the soil for the control of weeds and soil-borne pests and diseases, or are drilled into the soil as seed dressings. By these methods, potentially polluting chemicals are introduced into plants and the soil. Once chemicals are in the ecosystem their effect depends upon subsequent changes that occur. Some pesticides are described as stable and persistent, which means that they remain in the soil for varying periods of time. Gradually the quantity of persistent chemical builds up in the ecosystem as additional applications are made. Other pesticides are unstable or non-persistent, and they are quickly decomposed by soil micro-organisms carrying out a process called biodegradation. Animals absorb pesticides in their food, and some remains stored in specific body organs such as the liver, and is not excreted. Consequently as more of the chemical is absorbed, it accumulates within the body until it reaches a lethal threshold dosage for the species. The use of pesticides called organochlorines has caused the death of many predatory birds and mammals, which are consumers within food nets. The chemicals are transferred from plants or the soil to the different levels of food consumers. For example, birds eat seed grain treated with a fungicide dressing, or birds eat earthworms that have ingested contaminated humus derived from sprayed crops or trees, or foxes eat birds with body accumulations of insecticides derived from their food source. Cases are known where the use of pesticides has resulted in the elimination of an entire species population of birds and butterflies, in a specific region. The particular effects of pesticides are complex, and the simple examples that have been given are only to illustrate the fact that these chemicals can produce changes in food nets and in the numbers of species populations. In general terms, the harmful effects of pesticides can be directly related to their persistence and accumulation in the environment, as well as within the body tissues of animals and humans.

Insecticides

There are two main groups of pesticides that are in common use as insecticides. They are complex organic compounds, that are usually named by initials derived from their chemical structural name, or by proprietary names. The first groups are the organochlorines or chlorinated hydrocarbons, which have caused considerable pollution damage. They have four properties which cause them to be regarded as very toxic pollutants. All organochlorines are persistent and non-biodegradable in the environment. For example, if a field is sprayed with DDT it is found that up to 50% is still present in the soil after 10 years. This does not imply that the other 50% has been removed from the environment, but merely that it has been removed from the soil, perhaps into plants, the atmosphere, or water courses. Secondly, these insecticides have a wide range of biochemical action as plant and animal toxins, and they can cause changes in fertility, as well as in hormone and enzyme action. Organochlorines are also selectively soluble in fats, and so are retained and stored in the body of animals and humans in varying amounts for long periods. Lastly these chemicals are able to spread as wind blown dust, and in water, for long distances. DDT has now been detected in nearly all the world's rivers and oceans and even in the Arctic and Antarctic. It has been estimated that over 1 M tonnes of DDT has been used worldwide since 1947, so there is a considerable amount of this substance in the global environment. In the UK, very small quantities of organochlorines are present in drinking water, fruit, vegetables, and milk, with larger amounts in meat, fish, eggs, butter, and flour. In 1968 these quantities were of the order of 0.1 parts per million (ppm) or less. However UK estimations of DDT, and its breakdown product DDE, in human body fat in 1964 were 3.3 ppm on average, which shows how the dosage from ingested food can accumulate. In the US, body fat DDT levels as high as 12 ppm have been found, and this caused its use to be banned in 1972. Evidence of specific harmful effects in the human body are difficult to assess in the short term, but there is evidence of harmful longer term effects in experimental mammals. In the 1960s, evidence of the accumulation of organo-chlorines in the fatty tissues of humans and wild predatory species, together with the deaths of thousands of birds and mammals, caused the pollutive effects to be investigated. The Ministry of Agriculture, Fisheries and Food set up a working party which produced reports on toxic chemicals in 1951, on residues in food in 1953, and on risks to wild flora and fauna in 1955. These reports showed that organochlorines were causing harm to birds, fish, beneficial pollinating insects, and other forms of wild life. One chemical still causing concern is Dieldrin, which is still in common use in certain manufacturing industries for moth proofing and wood preserving, and this substance is very toxic and persistent. Another group of pollutive chlorophenol compounds, widely used in industry, are the polychlorinated biphenyls or PCBs. They are present in a number of products including lubricants, plasticizers, wax polishes, sealing compounds, and heat exchanger fluids, and they are released into the environment as a vapour or in solid wastes. PCB molecules are structurally allied to DDT, and they are toxic, persistent, and accumulate in body tissues. Some voluntary limitation in their use is now being operated by some manufacturers, but there is at present no control or overall agreement in force to restrict their wide ranging use.

The second group of synthesized organic insecticides are the organophosphorous compounds. They were developed during the 1950s, and are now being used in larger quantities than the more pollutive organochlorines. Organophosphorous compounds are unstable, they do not persist in the environment, and are quickly biodegraded in the soil. They kill pests quickly and effectively by attacking the nervous system, and causing inactivation of the enzyme acetylcholinesterase, which is responsible for controlling the transmission of nerve impulses. Unlike the organochlorines they do not accumulate in the body and so have less harmful effects upon food nets and humans. However some are highly toxic and present a hazard to anyone who uses them or ingests a liquid formulation, for example 28 g of TEPP could kill 500 people, and so great care is necessary in their storage and to prevent them entering water courses. Recently developed organophosphorus insecticides, such as Menazon, are selectively toxic and have a low toxicity to mammals. Carbaryl acts on insects in a similar way to the organophosphorus compounds, it may be used as a DDT substitute, and is non-persistent in the environment.

Fungicides

Most of the fungi that cause plant diseases are usually given common names describing their effects, such as mildews, blights, smuts, rusts, scab, spot, rot, etc. Fungi are pathogenic organisms that absorb all their nourishment from a host plant, and so many of them behave as parasites. Fungi may be present on foliage surfaces as powdery mildews, or they may ramify through plant tissues, and produce reproductive spores at the outer plant surface to enable the infection to spread. Fungicides are of two main types. The older surface fungicides such as Bordeaux mixture, and lime sulphur are applied to the foliage as sprays or dusts. The active ingredient does not appreciably penetrate the plant surface, but kills the fungus and the surface spores. The newer types of systemic fungicides are absorbed into the plant through the leaves, roots, or seeds, and are translocated within the plant tissues. Consequently the systemic types are more efficient than the surface fungicides. Both types of chemical compound must have a low toxicity to prevent damage to the plant crop tissues. The earlier types of fungicides often contained metals such as copper, zinc, mercury and arsenic, as well as sulphur and natural plant extractions. These were the only types of fungicides in use until the 1950s, but many types of synthesized organic substances have been brought into use during the last twenty years (Table 11.1).

There are a number of organic surface fungicides in use. Organic mercury fungicides are used chiefly as seed dressings for cereals, and for fruit scab diseases. The alkylmercury compounds are used in Sweden, and this has caused widespread mercury poisoning of birds and some mammals. Phenylmercury fungicides are used in the UK, but they appear to have caused little environmental damage. It is questionable whether mercury compounds should continue to be used, because it is a very toxic metal. The dire effects of discharging industrial effluents containing mercury into water, and its accumulation in food fishes are well known. The organic thiocarbamates, such as Thiram, Ziram, and Maneb, are replacing the use of mercury compounds for seed dressings, and for the control of grey mildews and damping-off disease in seedlings. Many phenolic substances are too toxic for plant use, but they are used as industrial fungicides for the protection of timber and textiles. However the ditrophenol compound, Dinocap or 'Karathane', is used for controlling powdery mildews on many horticultural crops. Systemic fungicides must have a low plant toxicity and be capable of being absorbed into plants, where they are converted into an active toxicant. Benomyl, and thiophanates are only two of a number of the new systemic fungicides that are being used for the control of a wide range of pathogenic fungi. Some organophosphorus compounds with a low mammalian toxicity, such as 'Dowco', and Triarimol are systemic fungicides used for the control of powdery mildews, and Carboxin is used as a seed dressing.

Fungi do not appear to mutate as readily as insects, so that few fungal strains have become resistant to surface fungicides. However the use of some systemic types has fairly quickly shown up resistant strains, especially where fungicides with a specific toxic action have been used. All fungicides have a low toxicity to plants, birds, and mammals, and so they are not potentially hazardous in the short term. However some products contain metals that are known to accumulate in food chains and nets, and humans in the long term period. The use of fungicides is increasing and particularly the use of organic compounds, so there is a need to research and monitor these pesticides in the future, to safeguard the environment.

11.3 Herbicides

Herbicides are a group of organic chemicals which are used to destroy or suppress the growth of plants. There are three types which can be classified according to their mode of action. Selective herbicides are used for destroying particular unwanted plants or weeds without injury to others, for example broad-leaved weeds growing amongst narrow-leaved cereals. Residual herbicides are applied to the soil at the time of seeding where they remain active for several weeks, so preventing the growth of weeds in competition with the emerging germinating crop. Translocated herbicides may be sprayed on to leaves and are able to move to all parts of the plant through the internal translocation mechanism. All these three types, if used before flower opening, will prevent seed formation and so reduce weed spread by seed dispersal. Some herbicides are called defoliants, and these are chemicals that cause leaf fall and the subsequent death of the plant. The increasing use of herbicides has developed since the 1940s alongside increased agricultural mechanization, the reduction of hand labour methods, and techniques to increase crop yields. Farmers use selective herbicides as pre-emergent soil treatment to destroy or prevent the growth of weeds, because these compete with crops for moisture, light, and nutrients, and so reduce yields. Horticulturalists and gardeners use herbicides as a labour saving way of removing unwanted weeds growing in crops, flower beds, and lawns. Non-selective or total herbicides are used as an easy way to destroy all vegetation on industrial sites, public paths, pavements, railway tracks, and roadside verges (see Table 11.1).

Herbicides have been used since about 1898, but the chemicals used before the 1940s were inorganic and rather crude and unselective in their action. For example copper (II) sulphate, sulphuric acid, and DNOC were all used for weed control in cereals, and sodium chlorate as a non-selective weedkiller. The newer organic herbicides can be classified into two groups according to their biochemical action upon plants. Chlorophenoxyl acid compounds such as MCPA, 2, 4-D, and 2, 4, 5-T have been called 'hormone weedkiller' or growth regulators, because they change the normal plant growth behaviour. They cause rapid, distorted, uncontrolled growth, and eventual death. The substance 2, 4, 5-T, as well as the persistent Picloram, were used as defoliants by the US forces in South Vietnam and Indochina in 1968–69. Excessive doses were spread by aerial spraying on the jungle vegetation and enemy crops. In South Vietnam 1400 km^2 of mangroves were treated, and by 1972 they had shown no sign of regeneration, so clearly there was considerable damage to the ecosystem. The other group of herbicides consist of triazones (Simazine), or carbamates, or bipyridyls (Paraquat and Diquat), and they inhibit photosynthesis. Their effect is to block vital stages in this process so that the plants die from lack of synthesized carbohydrate. Herbicides have a varying time of persistence in the soil, ranging from a few hours for Paraquat; a few weeks for MCPA, 2, 4-D and Dalapon; a few months for 2, 4, 5-T; and up to 1 year or more for Simazine and Picloram. When animals absorb them in food, they are generally not found to be toxic or to have cumulative effects, but their long term effects are not yet known.

The most important environmental effect of herbicides is the elimination of plants. All levels of consumers in food nets rely upon the basic producers which feed upon plants. Where wholesale defoliation is carried out the entire local ecosphere is devastated, and many insects, birds, and mammals are eliminated. There is concern about the possible long term effects of some herbicides on the human population. For example 2, 4, 5-T formulations usually contain small amounts of the highly toxic impurity called dioxin or TCDD, which is a known teratogen and liable to cause congenital malformations. Similarly 2, 4-D can affect bird embryos, and triazines may cause mutagenic effects in animals.

11.4 Pesticide Safeguards and Use

The use of pesticides appears to be a permanent part of man's scientific approach to the control of pests and diseases. They have been used by farmers, foresters, gardeners, housewives, and in industry, in all the Developed countries for the last 30 years. Now the Developing and the Third World countries are also using them in their efforts to improve food production to feed their increasing populations.

It has been realized, since the 1950s, that their use implies considerable hazard to wild life and the human population. Subsequently, various measures have been taken by the UK Government to safeguard the population and the environment.

In 1957 a voluntary Notification of Pesticide Scheme (NPS) was jointly agreed between the Government and manufacturers, whereby firms agreed to supply details of new pesticide products to the MAFF and other Departments, before they were put on sale. The NPS became the voluntary Pesticide Safety Precautions Scheme (PSPS) from 1964. The scheme aims to ensure that no product in use will be a hazard to the user, the general public, farm livestock, wild life, and domestic pets. A second voluntary scheme commenced in 1942 as the Crop Protection Products Approval Scheme, and subsequently became the Agricultural Chemicals Approval Scheme (ACAS) in 1960. This scheme provides for proprietary crop pesticides, which are given clearance under the PSPS, to be approved for their biological efficiency. The aim of the ACAS is to assist users in the selection of products of known performance for their specific control problems, and to discourage unsatisfactory products. Both the PSPS and the ACAS schemes are voluntary, and there is no specific UK legislation to control the use of pesticides. Some additional protection is afforded by the Farm and Garden Chemicals Act 1967 and the associated Regulations 1971, which require all pesticide products to be labelled, so as to show details of the active chemical ingredients present and a warning of any risks to people or animals when they are used.

Some progress has been made in trying to control the use of hazardous agro-chemicals in the UK. Heptachlor, Aldrin, and Dieldrin have not been used as spring sown seed dressings since 1962, and for autumn sown seed dressings since 1974. Heptachlor is now no longer available in the UK, and there are restrictions in the use of DDT. This organochlorine compound is not included in preparations sold for use in the home and garden. The use of these persistent pesticides has therefore declined, and between 1963 and 1972 their production was reduced by 47%, to 250 tonnes per year. However these hazardous substances are still being used in the UK, because it is affirmed that there is no economically viable alternative available. This statement particularly applies to DDT, which is still used on a world scale to control disease-carrying vector insects. The world usage of DDT is about 200 000 tonnes per year, of which 40 000 tonnes is used for the control of malaria. The continued use of this persistent compound is justified by the WHO on the grounds that it saves human lives.

Pesticides may be described as short term palliatives with potential long term human hazards. They may appear to be very effective in killing a particular pest at any one time, but the species is not eliminated for ever. Any species population is not homogeneous, but consists of individuals with genetic variability. Consequently the individual members of a population react to a pesticide chemical in different ways. When a new pesticide is brought into use some of the species population will be unaffected, or are said to be resistant. The majority will be susceptible and die, but the resistant members survive and continue to breed.

Eventually after numerous generations, the entire species population will be resistant to that pesticide, and it can no longer be used effectively. Because of this natural pest behaviour, the agrochemist is continually having to produce new types of effective pesticides. Alternatively it is common practice to use formulations which contain a mixture of different pesticides or herbicides, for example DDT plus Malathion or BHC, and 2, 4-D plus 2, 4, 5-T or MCPA or Mecoprop. This 'battery' method of use is an attempt to combat resistant species, but it increases the quantities of the various organic chemicals that enter the environment. There is substantial evidence to show that most organochlorines and some herbicides persist in the environment, and may have serious long term effects. A number of these synthesized compounds can produce carcinogenic effects in mammals, and it is suspected that long term effects in humans could be blood disorders, such as leukaemia or cancer of the red bone marrow.

Another objectionable aspect of the widespread use of insecticides and herbicides is that their action may be totally unselective. In an ecosystem, there are always some natural predators and parasites that have an ecological role in controlling the size of a species population. The use of unselective and persistent insecticides destroys both pests and predators, and so the natural biological control is disrupted. Herbicides may destroy all non-cultivated plants including wild flower species. This is causing the elimination of numerous harmless wild plants in fields and on road verges, and despoiling the countryside for naturalists and the general public. Herbicidal use also destroys the food plants of many insect species such as butterflies, and so is causing a reduction of the numbers of beneficial insects as well as pests. Therefore the question arises as to whether the wholesale use of chemicals is justified at all. The epidemiologists argue that insect vectors such as mosquitoes, flies, and fleas, which transmit human diseases, must be killed to improve human health conditions. Agriculturalists and horticulturalists argue that the use of pesticides is an integral part of modern agrarian technology. It is true that the use of these chemicals does help to improve the quality and yields of arable crops, vegetables, fruit, and flowers. Crop losses during 1971–72 have been estimated, and the results are given in Table 11.3. The table shows that the regions where there is little use of pesticides, such as Africa and Asia, did have significantly higher crop losses. Experiments over five states in the US have shown that the average crop yields were raised by up to 50% by the

Table 11.3. Crop losses during the winter of 1971–72

Region	Percentage loss caused by			
	Insects	Diseases	Weeds	Total
Europe	5.1	13.1	6.8	25
North and Central America	9.4	11.3	8.0	28.7
Africa	13.0	12.9	15.7	41.6
Asia	20.7	11.3	11.3	43.3
USSR and China	10.5	9.1	10.1	29.7

use of pesticides. To ban their use completely would be quite unacceptable to the industries concerned, and to the national economic policies of the Developed countries. However is it necessary to reduce the increasing amount of environmental pollution caused by organic chemicals. Organochlorines and other persistent compounds should no longer be used, especially as a number of alternative compounds are becoming available. The overall use of all other chemicals should be reduced. This could be achieved by using smaller treatments in a more selective manner, and only after the most rigorous laboratory and field testing of new formulations. Alternatively, and as a parallel development, new types of control methods need to be researched and developed to safeguard future human populations and the natural environment.

The ideal pesticide should have a high toxicity against a specific pest, it should only persist long enough to produce a high kill rate, it should not affect the rest of the ecosystem, and it should not harm natural predators and beneficial insects. Unfortunately the majority of pesticides in use at present have characteristics that are a long way short of the requirements of the ideal pesticide. Some recently introduced selective pesticides meet some of the ideal criteria, for example Menazon for aphid control, and Dimethirimol for use against cucumber powdery mildew. The increasing use of systemic rather than contact insecticides is encouraging, because their selective action ensures that only insects feeding upon the treated crop ingest the toxin and are killed.

Probably the best policy for the future is to concentrate on developing a joint biological and chemical approach to pest control. The concept of biological control of pests has been known for at least 50 years, and there are numerous examples of its successful use. In Australia, a prickly pear species spread over 24 hectares of land, and made it unusable for agricultural cultivation. In 1930, 3000 M eggs of an Argentine moth were introduced, and over the ensuing 7 years the moth larvae successfully destroyed the prickly pear plants. The vedalia 'lady' killer beetle has been used to control the cushion scale insect on Californian citrus trees. Other predators at various times have been used to control scale and mealy bugs, alfalfa aphids, and the gypsy moth and saw fly in conifer plantations. The infection of pests occurs naturally and this method of biological control has been investigated. There are bacteria, viruses, fungi, protozoa and microscopic worms that are known to parasitize insects and kill them. Fears that the use of these micro-organisms could endanger other species do not appear to be substantiated, because insect pathogens are very specific in their hosts and effects. Non-chemical methods of pest control are known to be feasible, and they could be extended and used to a greater extent in the future. However it is likely that their use on a large scale would be limited. Biological predators operate in their own ecosystem and they may not be successful if introduced into different environmental conditions. Also the practicality of raising millions of predators could prove to be impossible or uneconomic in relation to their control capabilities.

Most chemical pesticides in use aim to kill pests. An alternative method of insect pest control is to disrupt the normal biological behaviour. Many insects have a life cycle consisting of the four successive stages of egg, larva (grub or caterpillar), pupa (chrysalis), and adult insect. Some synthesized organic chemicals have been developed that will either prevent the life cycle from being completed, or affect adult reproduction and stop successive generations of insects developing. Synthesized growth hormones or regulators, such as diflubenzuron, have been used to prevent the growth and moulting of insect larvae, and their subsequent development into adults. This method of control has been used for livestock fly pests, some crop pests, and mosquitoes. Other synthesized growth hormones act to prevent the juvenile larval stage changing into the pupal stage. One of these so-called juvenile hormones is methoprene, and it has been used in the US for the control of public and animal health pests and mosquitoes. Other juvenile hormones act as chemosterilants and induce adult female insects to lay sterile eggs, for example. An alternative sex control method that has been used is the irradiation of male insects with X-rays or gamma rays. This produces sterility, and so after mating, the females lay unfertilized eggs that do not develop any further. The use of growth regulators, juvenile hormones and chemosterilants are the beginning of a new type of pest control. Compared to conventional pesticides the new types appear to have several advantages. They have a low toxicity to mammals and birds, are often effective against resistant species, and are highly specific and selective in their action. For example, the use of methoprene against mosquito larvae did not affect other types of aquatic insect larvae.

Clearly the pesticide control methods in operation over the last 30 years have produced environmental pollution. Assuming that some chemical pest control is still necessary, there needs to be a reduction of the types and quantities of pesticides used in the future. The non-selective persistent substances such as DDT must be phased out of use, because they are undeniably pollutants with a high potential human toxicity. The selective systemic pesticides do not seem to be excessively hazardous so far, but most of them have only been in use for a short time. Research and close monitoring of their effects needs to be carried out. Pesticide aims and methods should be fundamentally appraised. Chemical compounds should only be used if they have a highly specific action, are very effective, and need low concentrations, and a minimum number of applications for control. In addition, there needs to be much greater education of the public and the agricultural and horticultural industry regarding the use and environmental hazards of pesticides.

CHAPTER 12

Future Developments

Land pollution is basically caused by man's inability to safely and efficiently dispose of the waste materials resulting from all human and technological operations. The chief aspects of land pollution causing concern are solid and semi-solid waste disposal methods, the presence of hazardous chemicals in the environment, and the despoilation and degradation of the land surface. Chemical pollution is probably the most serious aspect, because substances can enter the natural cycles and ecosystems, and affect the food, health, and continued existence of animals and human beings. Since land is part of the ecosphere, land pollution does not exist in isolation. Some constituents of air pollution are deposited upon the land, and various chemical pollutants can move from the soil into streams, rivers, and the sea.

The annual amount of solid waste produced in the UK is not decreasing. So long as economic growth policies prevail in the future, there will be an increase in the quantity of waste produced. The central problem about waste disposal is how to contain it within the environment, and prevent pollution arising. Containment involves the use of land surface for deposition, and the type and volume of waste concerned. There is considerable competition for land use in many regions of the UK. Using land for tipping prevents it being utilized for useful purposes for a considerable time, even if it is eventually reclaimed. Future trends should be towards trying to reduce the volume of solid waste that needs to be tipped and occupy land. The largest amount of waste is produced by the mining and quarrying industries, and they should be encouraged to make more use of land in-filling and reclaiming re-usable materials, even if this means transporting them away from the waste production site. Local Authorities should also be pressed by Central Government to carry out more reclamation of materials from domestic and trade wastes. Also LAs should consider more use of incineration and pyrolysis for dealing with waste, and utilization of the heat energy released. The Control of Pollution Act 1974 provides the legislative powers to control waste disposal and it encourages the reclamation of useful materials by LAs. When this Act is fully implemented it is to be hoped that local disposal authorities will make full use of these powers, and considerably improve upon their own and industrial practices for waste disposal in the future. Pollution caused by hazardous chemical substances in the environment is apparently increasing for several reasons. More substances of many types, and particularly organic chemicals are being used, so that the possibility of their reaching harmful amounts is increasing. Research and medical studies are revealing hitherto unrecognised pollutive chemical effects in humans and animals, especially as long term effects become apparent. New organic compounds are continually being developed and introduced for industrial and agricultural use, and the medium and long term effects of these substances cannot easily be assessed. The monitoring of chemical concentrations in soil, water, food, and body tissues is not yet sufficiently co-ordinated and regular to provide early warning of hazardous levels. There can be no absolute standards or threshold limit values (TLVs) for chemicals. The determination of values can only be based upon the best evidence available at the time. Consequently TLVs have to be periodically reappraised as new pollutive chemical effects appear, and previously safe values may have to be designated as hazardous. The three main potential sources of chemical pollution on land are the dumping of solid wastes containing toxic substances; leachates from industrial spoil heaps; and land drainage containing agricultural fertilizers, silage effluent and pesticides. More toxic waste dumps will be needed in the future unless there is a radical change in the method of toxic waste disposal. The correct geological siting of these dumps is vital to prevent possible contamination of water supplies. The increased use of incineration and pyrolysis methods for toxic waste treatment, with adequate safeguards against air pollution, would substantially reduce the environmental hazards and the amount of land deposition required. The Control of Pollution Act 1974 should control the deposit of toxic wastes adequately, through the rigid implementation of the disposal licensing procedure. This needs to be complemented by regular monitoring of the adjacent land and water courses. Control over the use of chemicals in agriculture depends to a large extent upon the individual operators. This will probably continue to be an area of environmental concern in the near future. While accepting the need for increasing home food production, the farming techniques for achieving this should be reviewed and assessed in the future. There needs to be improved monitoring systems for the soil, water courses, and foodstuffs, to detect increasing chemical concentrations. This should be accompanied by increased research into the medium and long term effects of agricultural chemicals upon wild life and human health. Control and use of radioactive substances is provided by legislation and monitoring procedures. The TLVs must be constantly reviewed on an international scale. The chief radiological environmental problem of the future is the disposal of wastes using techniques which prevent the possibility of radioactive pollution to future generations. In the long term, beyond the present century, the use of uranium and plutonium for nuclear fission reactors may hopefully be an interim stage in the development of nuclear energy. The ultimate process of energy production for succeeding centuries is nuclear fusion.

The third aspect of land pollution is concerned with the condition of the land surface, and this can be variously described as degradation, or despoilation, or dereliction of the environment. The approach to this type of land pollution is quite different to the other two aspects, because it does not directly affect the health of the population. People are becoming increasingly concerned about the degradation of the environment because it reduces the 'quality of life'. This phrase has many connotations, but it includes desirable living conditions where there is clean air, clean water, uncontaminated food, and enjoyment of the countryside, wild life, recreational pursuits, aesthetically pleasant residential surroundings, etc. Anything that affects the maintenance of these conditions can be considered as pollution, and degrades the quality of life. The recognition of this type of pollution is often a matter of subjective judgement and attitude, and not qualitative observation and quantitative measurement. It is not surprising therefore that because of its relatively low impact upon the population, this type of land pollution has received little priority or finance in the past. To effect an improvement in the future will require basic changes of attitude and policy at both local and national levels. For example, it needs to be acknowledged that derelict land is totally unproductive, it has a derogatory social and aesthetic effect upon the local inhabitants, and can inhibit new developments. Most rural and inner city derelict land can be reclaimed for agricultural use, or redeveloped as part of an industrial or residential community. Local Planning Authorities should take a more responsible role in exercising their powers of planning control. Environmental improvements can be made in a number of ways. When the development of new roads and motorways, industrial estates, mining and quarrying operations, and solid waste tips are planned, the total pollutive effects upon the local neighbourhood should be more fully considered and recognized. The transportation of heavy goods and hazardous chemicals by road needs assessment and change. The volume of heavy goods traffic on roads is rapidly increasing, and so are accidental spillages of loads causing air, land, water pollution, and hazards to health. The alternative of rail transport should be reassessed, particularly for bulk chemical and radioactive waste transport. The need to conserve countryside amenities of all types will probably have to be more fully recognized in the future. Predicted future employment trends indicate a shorter working week, which implies that workers will be spending more leisure time in the non-productive environment of their homes, recreational and rural areas. Therefore it would seem necessary to use more national resources for maintaining and improving the non-working environment. Also the views of the general public need to change, so that the present apathetic attitude towards the degradation of the environment is replaced by an improved social responsibility and aesthetic appreciation. Then the destruction of rural amenities, the scattering of litter, and the indiscriminate dumping of rubbish will diminish. All the above suggested changes and development trends involve spending money, much of which must be provided by Central Government. At present some types of land pollution are receiving attention, but changes are taking place in a very piecemeal manner. So far no government has been able to produce an overall comprehensive policy for the future containment and reduction of land pollution. Unfortunately the present public apathy towards land conservation and the redemption of derelict land provides no stimulus to governments to change their policies.

References

Arvill, R., *Man and the Environment*, Penguin, 1973.
Atkins, M. H., and Lowe, J. F., *Pollution Control Costs in Industry*, Pergamon Press, 1977.
Attenborough, Pollitt, and Porteous, (Eds), *Pollution: the Professionals and the Public*, Oxford University Press, 1977.
Barr, J., *Derelict Britain,* Penguin, 1970.
Besselievre, E. B., *The Treatment of Industrial Wastes*, McGraw-Hill, 1969.
Brooks, P. F., *Problems of the Environment*, Harrap, 1974.
Carson, R., *Silent Spring,* Penguin, 1965.
Civic Trust, *Derelict Land*, Civic Trust, 1967.
Cremlyn, R., *Pesticides*, John Wiley, 1978.
Department of the Environment, Building Research Establishment, *Uses of Major Industrial By-products and Waste Materials,* BRS, 1974.
Department of the Environment, *Disposal of Awkward Household Wastes*, HMSO, 1974.
Department of the Environment, *Pollution: Nuisance or Nemesis*, HMSO, 1972.
Department of the Environment, *Refuse Disposal*, HMSO, 1971.
Department of the Environment, *Taken for Granted—Report on Sewage Disposal*, HMSO, 1970.
Department of the Environment, *War on Waste—A Policy for Reclamation*, HMSO, 1974.
Diamant, R. M. E., *The Prevention of Pollution,* Pitman, 1974.
Ehrlich, P. R., and Ehrlich, A. H., *Population, Resources and Environment*, W. H. Freeman, 1972.
Ehrlich, P. R., Ehrlich, A. H., and Holdren, J. P., *Human Ecology*, W. H. Freeman, 1973.
Escritt, L. B., *Public Health Engineering Practice*, Vol. 2, 4th Edn., McDonald and Evans, 1972.
Fletcher, W. W., *The Pest War*, Blackwell, 1974.
Gilpin, A., *Dictionary of Environmental Terms*, Routledge and Kegan Paul, 1976.
Holister, G., and Porteous, A., *The Environment: A Dictionary of the World Around Us*, Arrow, 1976.
Mellanby, K., *Pesticides and Pollution,* Collins, 1972.
Mellanby, K., *The Biology of Pollution: Institute of Biology Study No. 38*, Arnold, 1972.
Open University, *Maintaining the Environment, T100 Units 26 and 27*, O. U. Press, 1972.
Open University, *Municipal Refuse Disposal and Toxic Wastes, PT272 Units 9 and 10*, O. U. Press, 1975.
Overman, M., *Water*, Aldus Books, 1968.
Revelle, C., and Revelle, P., *Source Book of the Environment*, Houghton Mifflin, 1974.
Royal Commission on Environmental Pollution, *First Report*, HMSO, 1971.
Royal Commission on Environmental Pollution, *Second Report*, HMSO, 1972.
Royal Commission on Environmental Pollution, *Third Report*, HMSO, 1972.

Royal Commission on Environmental Pollution, *Fourth Report*, HMSO, 1974.
Royal Commission on Environmental Pollution, *Seventh Report*, HMSO, 1979.
Skitt, J., *Disposal of Refuse and Other Waste*, Charles Knight, 1972.
Tearle, K., (Ed.), *Industrial Pollution Control*, Business Books, 1973.
Whitby, Robbins, Tansey, and Willis, *Rural Resource Development*, Methuen, 1974.
Yapp, W. B., *Production, Pollution, Protection*, Wykeham Publications, 1972.

SECTION IV

Water

CHAPTER 13

Water Supplies

13.1 Water consumption

Water is a basic natural resource required by all human beings, and by the modern technological society in which they live. Man requires a minimum body intake of water that varies from 2.8 to 13 litres per head per day depending upon the climate and the temperature. Water is normally taken into the body in food and drink, and the intake must balance the body loss resulting from breathing, sweating, and the excretion of urine and faeces. If there is no intake of water into the body, death can ensue within 10 days. Water is also essential to man for maintaining personal body hygiene and freedom from disease.

In addition to personal use, water is required for many other purposes, and these can be classified into five main user categories.

1. Public water supplies provided by ten Regional Water Authorities (WAs) in England and Wales. Two thirds of these supplies are used by domestic households for drinking, cooking, dishwashing, general cleaning, laundering, personal washing and bathing, lavatory flushing, car washing, and garden watering. The other one third is used by industry, and commercial and trade premises.

2. Industrial water supplies provided by the WAs or obtained by direct abstraction. Various industrial processes require large quantities of water for cooling purposes, steam raising, material processing, and the disposal of waste. Water is also used as a fluid carrier for processing materials such as paper fibres, or crushed ore.

3. Cooling water is used in large quantities for the generation of electricity by the Central Electricity Generating Board (CEGB) power stations. This water is usually abstracted from rivers, lakes, and estuaries.

4. The agricultural industry uses comparatively small quantities of water for dairy processing, animal hygiene, stock watering, and land irrigation. The horticultural industry uses water for land irrigation, glasshouse watering, and washing marketable vegetable crops.

5. Water is also required for amenity and recreational purposes. This category differs from the previous four because water is not abstracted from the hydrological cycle. The so-called water space includes streams, rivers, reservoirs, estuaries, canals, and coastal waters. It is used for all types of water sports, such as swimming, fishing, boating, sailing, ski-ing, and as a means of transportation

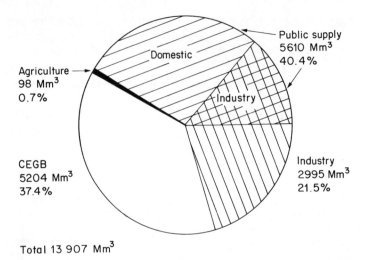

Total 13 907 Mm^3

Figure 13.1. Annual water abstraction in England and Wales 1975
(data from DOE Water Unit).

for pleasure or commercial purposes. The presence of areas of water also adds to the amenity value of the countryside and the enjoyment of the general public.

The amount of water used is usually described as water consumption. All water abstracted or drawn from the hydrological cycle is eventually returned to it after use. Therefore the term water consumption means the quantity of water abstracted for any purpose, irrespective of the condition or the time in which it is returned to the source or the atmosphere. The amount of water abstracted in England and Wales in 1975 is shown in Figure 13.1.

Water quantities are normally expressed in litres or cubic metres. One cubic metre (m^3) = 1000 litres = 220 Imperial gallons. For Figure 13.1 the comparable consumption figures per day were: public supplies 15.4 Mm^3, CEGB 14.3 Mm^3, industry 8.2 Mm^3, and agriculture 0.27 Mm^3, giving a total of 38.17 Mm^3. Figure 13.1 shows that 40.4% of the total water abstracted was used for public water supplies, and one third of this was used by industry; the remaining 3740 Mm^3, or 32.5% of the total consumption, was supplied to domestic premises, and is equivalent to a per capita consumption of 205 litres per day. Of this total, 29% is used for lavatory flushing and waste disposal, and 27% for personal washing and bathing. The CEGB is the next largest consumer, using 37.4%, but most of this water is quickly returned to its source after cooling use in power stations. Other industries consumed 4865 Mm^3, or 35% of the total abstraction. The amount of water required by different industrial processes varies considerably, for example 1.5 m^3 per tonne of product for coal, 273 m^3 for paper, and 900 m^3 for ammonium sulphate production. The total per capita consumption, including all types of users, was 307 litres per day in 1975. This compares to 251 l per day in 1965, and 289 l per day in 1970.

The total water usage in GB is rising by 1–2% per year. The increasing demand is caused by a number of factors. There has been a considerable population increase that has created a need for more houses and expanded industrial production. The British standard of living has improved so that more houses have flushing toilets, baths and showers, washing machines, dishwashers, cars to be washed, gardens to be watered, etc. Water consumption is expected to continue increasing in the future. The Water Resources Board in 1971–2 estimated that the public water supply demand in England and Wales would double from 14.3 Mm^3 per day in 1971, to 28 Mm^3 per day in the year 2001. This future estimate is based upon the assumption that the population will increase from 48.6 M to 57 M, and the domestic per capita consumption will increase from 295 1 to 490 1 per day, with an annual increase of 1.8%. Whether an increase of this scale will occur is arguable, but clearly the sources of supply and the amount of water available are of crucial importance for the future. As pollution can affect the quality and quantity of the water abstracted for use, then this must also be an important consideration.

13.2 Sources of Water

The input to the hydrological cycle on land is precipitated water falling as rain, hail, or snow. When this water reaches the ground some of it is evaporated, or runs off into the sea, streams, and rivers, and the remainder enters the ground water system (see Figure 2.1). Man obtains water supplies by abstraction from the hydrological cycle, which provides two chief sources of supply. Surface water is drawn from streams, rivers, lakes and storage reservoirs, and ground water is abstracted from underground aquifers through bore holes and wells. The proportion of water abstracted from these two sources, by the different user categories, is shown in Figures 13.2 and 13.3.

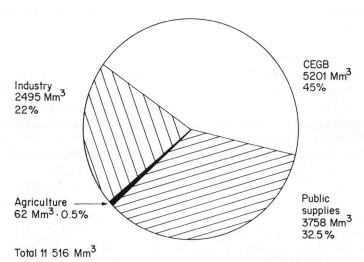

Industry
2495 Mm^3
22%

CEGB
5201 Mm^3
45%

Agriculture
62 Mm^3 · 0.5%

Public
supplies
3758 Mm^3
32.5%

Total 11 516 Mm^3

Figure 13.2. Surface water abstracted in England and Wales 1975
(data from DOE Water Unit).

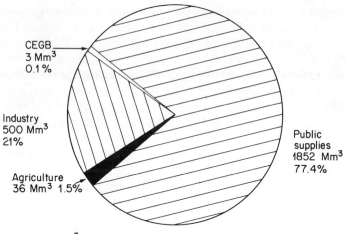

Total 2391 Mm³

Figure 13.3. Ground water abstracted in England and Wales 1975
(data from DOE Water Unit).

Figures 13.2 and 13.3 show that the total water abstraction in England and Wales, from both sources, was 13 907 Mm³ in 1975. The quantities consumed by the various user categories were 5610 Mm³ or 40% for public water supplies, 5204 Mm³ or 37% for CEGB power stations, 2995 Mm³ or 22% for other industries, and only 98 Mm³ or 0.7% for agriculture. Surface water provided 83% of the total consumption, and 17% was drawn from ground water aquifers. Comparative use of the two sources shows that industry obtained 83% from surface waters, the CEGB 99.9%, and public water supplies 67%. So surface waters provided the main water supplies for all types of users. The quality of surface waters, the degree of pollution present, and fluctuations in the volume of water available are obviously crucial factors that determine the volume of water supplies available for all users.

In GB, the long term average rainfall is 1050 mm per year, and this should theoretically satisfy all water requirements. But the rainfall is unequally distributed across the various regions of the country. For example, the average annual rainfall in the north and west is over 2500 mm, but in the south-east it is less than 500 mm. The areas of high-density population and industry are Tyne-Tees, Lancashire, Yorkshire, the Midlands, South-east Wales, London and the Home Counties. These areas have a comparatively high water consumption, and most of them are located where there will be an estimated water supply deficiency before the year 2000 (see Figure 13.4). Therefore there are likely to be problems of water supply to satisfy the future consumption in these areas. In addition to the medium term supply situation, there is the short term position involving fluctuations of supply over a 12 month climatic season. Normally there is a higher rainfall in winter than in summer, and this causes seasonal variations in the flow of rivers that provide a major source of water. These variations in rainfall and area consumption present problems of water distribution and

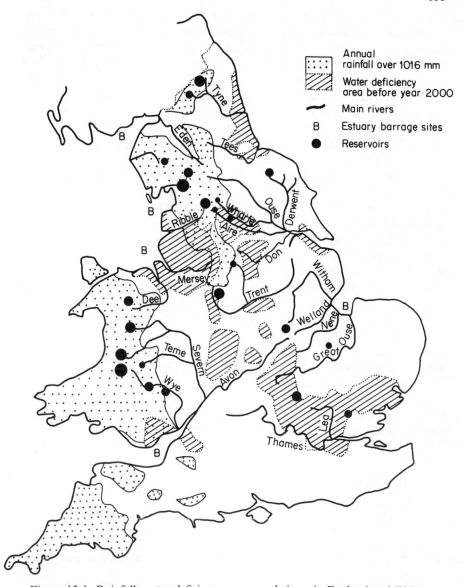

Figure 13.4. Rainfall, water deficiency areas, and rivers in England and Wales.

supply. Clearly the largest source of water occurs in the west and north, but the highest demand is chiefly in the central and south-eastern areas of England. The Regional Water Authorities have a duty to control water resources and supply,

13.3 Water Abstraction

Rivers may be considered to have three functions. They provide a source of water for all categories of user, but at the same time they are used for the removal of effluent discharges from industrial plants and sewage works. Rivers also have a recreational and amenity role in providing facilities for fishing, swimming, boating, and sailing in the countryside and towns. To maintain the third function rivers must have a balanced ecosystem that can sustain biological food chains and the natural cycles. If the river ecosystem is disrupted by pollution, then both the recreational and amenity functions are diminished, because organisms die and putrefying and stagnant conditions occur.

Water abstractions from rivers should not affect the other two functions. The annual flow of a river may fluctuate between a swollen flood condition and a minimum dry weather flow. Figure 13.5 shows the variations of flow occurring at one station on the River Wye. The volume flow is above the monthly average between November and March, and below it during the period April to October, and this is typical of a British river. The volume of river water throughout the year should be maintained at some pre-determined rate of flow, if the river is to fulfil the three functions mentioned and be free of pollution. For example if the river is capable of receiving effluent discharges without causing pollution, then there must be sufficient water flow to dilute the effluent volume eight times to conform to the commonly accepted Royal Commission 1912 standards. If the dry weather flow is too low, or the volume of water being abstracted is too high, then river pollution can occur, especially in the dry summer period. Clearly the volume flow of the river and the amount of pollution present in the water are limiting factors affecting the amount of water that can be abstracted at any one location at any one time.

To try to ensure continuity of an adequate water supply, the WAs make use of stored water in reservoirs and lakes, which can be used in two ways. The stored water can be used to provide a separate direct supply of water, via an underground pipeline or aqueduct, to augment fluctuating river supplies. For

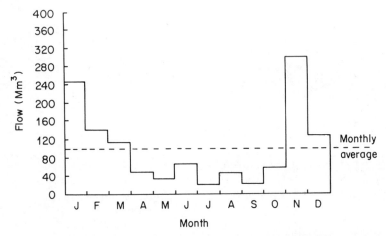

Figure 13.5. Total monthly volume flow in river Wye at Belmont, near Hereford, 1970/71. (Total discharge 1226 Mm^3.)

example Haweswater and Thirlmere in the Lake District supply water to Manchester, and Birmingham receives water from reservoirs in the Elan Valley in mid-Wales. Alternatively, newer types of reservoirs are being used to supply water directly into a river to regulate and maintain the water flow above a pre-determined level. These regulating reservoirs are in use for the rivers Dee, Severn, Tees, and Towy. For example, the Clywedog reservoir at Gwyneda in North Wales can supply 0.52 Mm^3 per day into the River Severn if required. The reduced summer flow of rivers can also be assisted by the use of storage basins in enclosed areas of low lying land. These are in use to augment the rivers Thames, Nene, and Welland.

Aquifers or water-bearing rock layers can be used as an alternative to surface water supplies. Aquifers provide a natural storage for ground water, and they exist in rock strata such as sandstone, some limestones, greensand and chalk, sited above impermeable rock. Water is abstracted by wells or bore holes, and it is replenished by ground water infiltrating from around the aquifer. The occurrence of aquifers is limited by the type of rock strata, so they are mainly located south-east of a line joining Newcastle upon Tyne and Torquay. Ground water from aquifers provides about half the total water consumption for this area, and constitutes about 16% of the licensed water abstraction in England and Wales. Aquifers can become polluted in two ways. If they are sited in coastal areas where the porous rock is below sea level, there may be an intrusion of sea water, when the level of fresh water falls below a critical point in the aquifer. Pollution can also occur if an aquifer is supplied by ground water that has percolated through polluted ground or a land tip. The use of aquifers for water supplies is limited by the rate at which they are recharged from rainfall. Recently the experimental

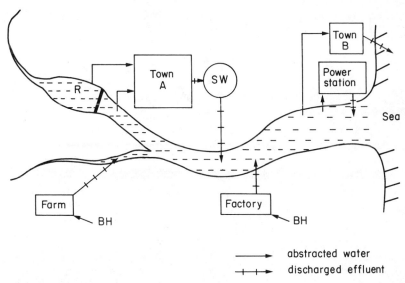

Figure 13.6. Abstractions and discharges in a river catchment area: BH, bore hole; SW, sewage works; R, reservoir.

recharging of aquifers by recharge bore holes has been carried out, using water from storage basins or water mains supplies. Again care must be taken to ensure that only non-polluted recharge water is used.

In any one district, such as a river catchment area, the water supplies may be abstracted from several different sources. In Figure 13.6 bore holes supply the isolated farm and the factory situated south of the river. The river headwaters have a storage reservoir that supplies water to the industrial town A, and river abstractions provide water to the inland town A, the coastal town B, and power station. Effluent is discharged into the river from the farm, southern factory, sewage works and power station. The coastal town discharges untreated effluent into the in-shore waters.

Figure 13.7 shows the catchment area of the River Trent, which contains six major rivers, and nine large towns. The area had a population of 2.18 M people in 1971, and a water consumption of 1.562 Mm^3 per day. The water courses supply water to this highly industrialized area and receive the discharged effluent produced by the population and industrial installations. In 1972 the DOE River Pollution Survey of England and Wales showed that the 2344 km (1457 miles) of

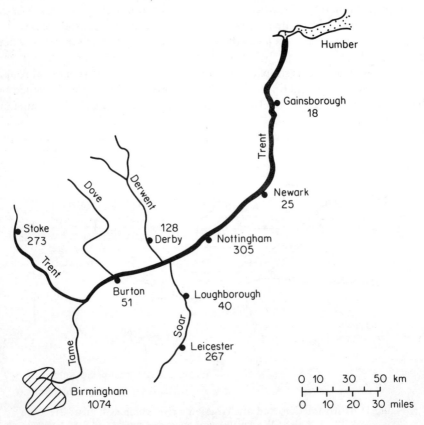

Figure 13.7. The Trent catchment area. For each town, the population is shown to the nearest thousand.

non-tidal river in the Trent River Authority area had a classified river quality of 42.4% unpolluted, 36.8% doubtful, 10.8% poor, and 10.1% grossly polluted. This River Trent study shows how the problems of water pollution impinge upon the surface water supply from rivers.

In 1975, 83% of the total water abstraction in England and Wales was obtained from surface waters, and water consumption is rising each year. There is growing public resistance to the construction of more upland reservoirs, which already occupied 200 km^2 of land surface in 1973. The increasing mobility of the public facilitated by cars and motorways is causing an increasing demand for the maintainence and conservation of upland leisure and recreational areas. An alternative to providing more storage reservoirs to meet increasing water consumption is to make better use of surface water resources. This can be achieved by reducing waste and pollutive discharges into rivers and canals, and increasing the abstraction from these waters at various points along their length. Plans are already in operation by the WAs to establish a type of water grid using interlinked rivers and canals, to achieve a better distribution of the available surface water.

13.4 Public Water Supply Treatment

In GB, all public water supplies receive a varying amount of treatment before entering the mains water supply. The 'raw' water is abstracted from natural waters, which may be polluted to a varying extent. The treatment of water supplies is important therefore, and it is relevant to know the details of treatment and the standards of public water supply that are required. There are five main aims of water supply treatment:

(1) to reduce the total micro-organism content, and remove bacteria and viruses that are harmful to health;
(2) to remove chemicals that are harmful to health;
(3) to remove suspended matter;
(4) to reduce to a low level those chemicals that might interfere with the normal domestic and industrial uses of water;
(5) to remove any corrosive properties of water, and protect the piped distribution system.

The Water Authorities are obliged to supply water that is clean, colourless, and free from disagreeable taste and odour. Note that there is no obligatory requirement regarding the chemical content of public water supplies in GB, but the World Health Organisation (WHO) recommendations are used by the WAs as a guide. These are shown in Table 16.1.

The impurities present in water may be grouped into five types, namely floating and suspended solids of visible size; small suspended and colloidal solids causing turbidity; micro-organisms; dissolved solids producing hardness and acidity; and dissolved gases such as oxygen, carbon dioxide, and hydrogen sulphide. Water treatment aims to remove most of these impurities, and the methods used are related to the particle size and chemical composition. The type of surface water treatment is not stardardized across different Water Authorities, and it

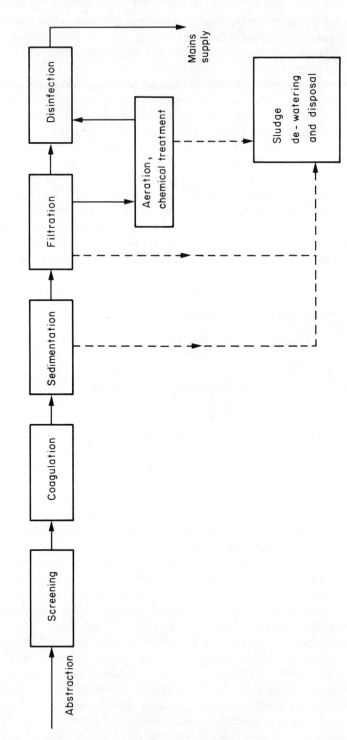

Figure 13.8. Flow diagram of a water treatment plant.

varies according to the raw water quality and the extent of water purification that is required. Complete treatment of water supplies may involve the five main stages of preliminary screening, coagulation, settling, filtration, chemical content treatment, and disinfection, as shown in Figure 13.8. Suspended solids and the larger micro-organisms are filtered out, by passing the water through a series of screens ranging down to micro-straining size. Suspended colloidal particles, most of which carry a negative electric change, need special treatment to induce aggregation and settlement. This is carried out by the addition of chemical coagulants such as organic polyelectrolytes, lime, iron salts, or PAM (polyacrylamide). Screening and coagulation will remove up to 90% of the suspended matter which settles out in the sedimentation stage. To remove the remaining turbidity, filtration is used, involving slow sand and rapid gravity methods. In some water treatment plants further processing is concerned with the dissolved chemical content, but in others this stage is omitted depending upon the source of the water. Chemical treatment involves the removal of hardness salts, iron, and manganese which affect water quality and taste, further removal of organic matter, adjustment of the acidity or pH value, and reduction of the corrosive properties of the water. Treatment may include chemical precipitation, and/or aeration which induces biological action upon the organic content, as well as displacing dissolved carbon dioxide to raise the pH value. Finally the treated water is disinfected to kill any harmful bacteria and viruses, and this is usually carried out by multi-stage chlorination or ozone treatment. The impurities that have been removed during treatment accumulate as sludge, and this has to be disposed of on to land or is dumped at sea. Even if all the five stages of treatment are carried out, the water still contains dissolved chemical substances. Water treatment is unlikely to alter the quantities of magnesium, fluorine, aluminium, zinc, copper, lead, nitrite, nitrate, and chloride that were present in the original raw water.

Ground water from aquifers is also a source of public water supply, and this is usually of a higher quality than surface water and less likely to be polluted. Some water treatment is often given, but this may only consist of disinfection, the removal of hardness salts and iron, and the adjustment of pH.

Water treatment is expensive in terms of the capital cost of plant and its operation. A 1970 estimate of the capital charge per unit volume of water treated was 0.25 pence per m^3, plus a minimum of about 1.5 p for treatment charges. Treatment of raw water is very necessary to safeguard the health of the population, and it will become more important in the future. Treatment is necessary because of the poor quality of many surface waters, and this is due to the discharge of polluting waste effluents into rivers. Water demand is increasing and this is necessitating more recycling or re-use of river water, which may be considerably polluted. These two factors will cause an increased water treatment facility to be required in the future, and so raise the treatment costs. In addition, to maintain the required water supply quality standards, more advanced methods will be needed to control certain pollutants that are not affected by the present treatment methods.

CHAPTER 14

Waste Disposal and Water Pollution

14.1 The Extent of Water Pollution

Water pollution may be defined in various ways. To the general public it is usually evident in terms of the observed appearance, so that a river is polluted if it is turbid, has foam on the surface, has an objectionable smell, and does not support fish and other living organisms. Another view is that water pollution adversely affects the aquatic ecosystem in terms of the living organisms, oxygen content, the presence of toxins, and so on. One type of definition relates pollution to the effects caused by man's activities: for example, 'A river may be said to be polluted when the water in it is altered in composition or condition, directly or indirectly as a result of the activities of man, so that it is less suitable for all or any of the purposes for which it would be suitable in its natural state'. Alternatively the pollution effects may be more specifically stated, as for example in this definition: 'Pollution is a natural or induced change in the quality of water which renders it unusable or dangerous as regards food, human and animal health, industry, agriculture, fishing, or leisure pursuits'. Basically pollution is induced by those human activities which cause pollutants to enter natural waters. Therefore another approach is to define a pollutant as 'a substance or effect which adversely changes the environment by changing the growth rate of species, interferes with food chains, is toxic, or interferes with the health and amenities of people'.

Despite the increasing amount of water pollution occurring during this century, no national survey to determine the extent of the problem was carried out until 1958. In this year, the then Ministry of Housing and Local Government coordinated an informal survey in England and Wales of all tidal and non-tidal rivers and canals that had a summer dry water flow of at least 4545 m^3 per day. The survey classified the water courses according to the nature of the discharges they receive, the oxygen content, biological oxygen demand (BOD), and the biological condition. Four classes were identified, as follows:

Class 1. Rivers and canals that were unpolluted, with a BOD below 3 mg/l.

Class 2. Rivers and canals of doubtful condition, needing improvement, and known to receive toxic and/or turbid discharges.

Class 3. Rivers and canals of poor condition and urgently needing improvement, with a dissolved oxygen saturation below 50%.

Class 4.   Rivers and canals that were grossly polluted, unable to support fish life, and were offensive in respect of odour and appearance, with a BOD of 12 mg/1 or more.

Class 1 and 2 waters are suitable as a source of public water supply, class 3 are only useful for irrigation and industrial cooling water purposes, and class 4 waters are unsuitable for any purpose. Following the success of the pilot survey in 1958, subsequent surveys were carried out and coordinated by the Department of the Environment in 1970, 1972, and 1975.

Table 14.1.   River surveys in England and Wales, 1958 to 1975

Year	Type of water	Length (km)	Class (%)				
			1	2	3	4	3 + 4
1958	All rivers	33 277	73.7	11.8	7.3	7.2	14.5
	Canals	2 608	58.0	25.0	9.0	8.0	17.0
1970	All rivers	38 566	74.1	15.4	5.7	4.8	10.5
	Canals	2 464	45.4	39.1	8.8	6.7	15.5
1972	All rivers	38 560	75.3	15.3	4.9	4.4	9.3
	Canals	2 472	45.7	39.7	9.5	5.1	14.6
1975	All rivers	41 420	74.0	17.3	4.9	3.8	8.7
	Canals	2 413	50.7	38.3	7.3	3.7	11.0

Table 14.1 shows the overall extent of pollution, and Figure 14.1 illustrates the separate conditions that exist in tidal and non-tidal rivers and canals. The four surveys are comparable because they were made over approximately the same length of water courses. The surveys may underestimate the true state of pollution as they only show the percentage length of polluted water in each class. The most important criterion is the volume of flow that is polluted, because this is the potential water supply that is available. For example, the river Cam at Cambridge has a mean summer flow of 136 360 m^3 per day, but the river Thames at Teddington has a flow of 2.86 M m^3 per day. The overall situation shows that some progress has been made towards reducing the amount of pollution in England and Wales. Between 1958 and 1975 the percentage length of class 3 and 4 rivers has reduced by 40%, from 14.4 to 8.7%. The polluted state of canals has also been reduced by 35%. However this apparent improvement is not uniform over the whole country, and some river catchment areas such as the Trent and Thames have improved, whilst others have deteriorated. Between 1970 and 1972, for every 11.3 km of river that improved, there was 1.6 km that had deteriorated.

A comparison of the three types of water courses shows differences in the extent of pollution. If the class 3 and 4 waters are expressed as a percentage of their total length, then in 1972, the quantity of severely polluted water in non-tidal rivers is 7.9%, in tidal rivers it is 27.4% and in canals it is 14.6%. Clearly tidal rivers are in the worst condition, because they receive pollutive discharges from the upper non-tidal reaches as well as the direct estuarine discharges. The most heavily polluted rivers are the Tyne, Tees, Wear, Trent, Humber, and Mersey,

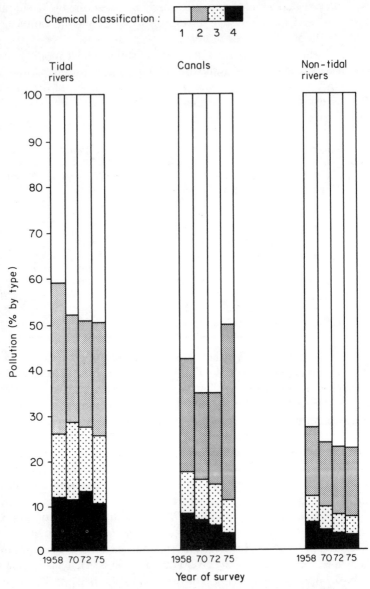

Chemical classification:

1 2 3 4

Figure 14.1. River pollution in England and Wales 1958–1975.

which receive the discharged waste from millions of population, and a heavy concentration of manufacturing industry. The problems of effluent disposal can be appreciated by a study of one river such as the Mersey (Figure 14.2).

The extent of water pollution may be expressed in terms of the standard water treatment test to determine the BOD value. This is an indication of the amount of biodegradable organic material present in the water, but it does not directly

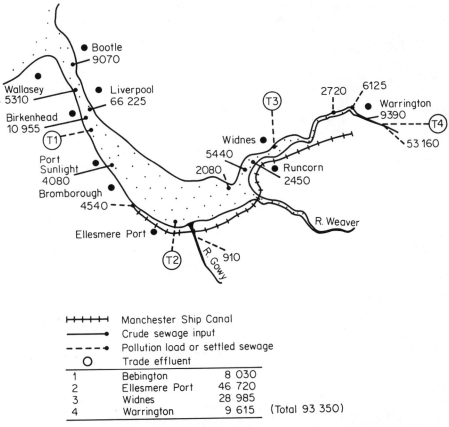

Figure 14.2. The Mersey Estuary pollution load, 1971 (adapted from the Royal Commission on Environmental Pollution *3rd Report* (HMSO, 1972)). The figures show pollution load as kg BOD per day.

indicate the amount of other non-organic pollutants. The data shown in Figure 14.2, as kg BOD per day, are therefore an indication of the amount of potentially polluting material that is discharged into the river.

The information shown is derived from the Mersey and Weaver River Authority and indicates that there is a daily sewage discharge of 182 455 kg BOD or 66% of total, and 93 350 kg BOD or 34%, of trade effluent. A considerable number of discharges contained untreated sewage, and these amounted to 62% of the total discharge load of 275 800 kg BOD per day. The type of treatment given to the trade effluent varies considerably, but much of this would certainly contain highly pollutive chemical material. The Mersey estuary receives non-tidal river water from the rivers Mersey, Weaver, Gowy and other tributaries, and this was classified in 1969 (Table 14.2) The 1972 DOE river survey supplied the information in Table 14.3. These tables show that 34.7% of the catchment river length was urgently in need of improvement, and only 34.9% was relatively

Table 14.2. Classified individual discharges to non-tidal waters (adapted from *Mersey and Weaver River Board Annual Report*, 1969)

Discharges	% Usually satisfactory	% Border-line	% Usually poor or bad	No. of discharges over 1000 litres per day
Sewage works	43.5	26	30.5	255
Trade premises	21.9	20	58.1	178
Total	34.6	23.3	42.1	433

Table 14.3. River pollution survey—Rivers Mersey and Weaver, 1972

Class	Percentage non-tidal river	Percentage tidal river	Percentage all types of river
1	37.5	1.5	34.9
2	32.5	4.5	30.4
3	10.5	59.7	14.0
4	19.6	34.3	20.7
River length	1400 km	107 km	

unpolluted. The non-tidal waters received 42.1% of very unsatisfactory or badly polluted effluents, and 31.1% of the river length was urgently in need of improvement. The tidal river was much more heavily polluted to the extent of 94% of its length. The River Mersey is only one of six heavily polluted large rivers in England and Wales, and this study illustrates the considerable problems associated with waste disposal from a large urban and industrial area.

14.2 Types of Effluents

The main cause of water pollution is the discharge of solid or liquid waste products containing pollutants on to the land surface, or into surface or coastal waters. The chief types, and the methods of disposal of waste products are summarized in Figure 14.3. The wastes that contribute towards water pollution may be broadly grouped into sewage, industrial, and agricultural types. Sewage is literally the contents of sewers, and these comprise the sewerage system that carries the water-borne wastes of a community. Sewage originates from domestic and commercial premises, land drains, some industrial plants, and agricultural sites. Other industrial waste is discharged directly into rivers, canals, and the sea, and not into the sewerage system. Industrial cooling water is mainly discharged from Central Electricity Generating Board (CEGB) power stations, but some comes from industrial plants. Cooling water has a low chemical content, but it is classified as a pollutant because the raised temperature at the point of discharge may cause thermal water pollution. Industrial waste is legally called trade waste, and this is defined in the Control of Pollution Act 1974 as 'any liquid, either with

Graph on page 144

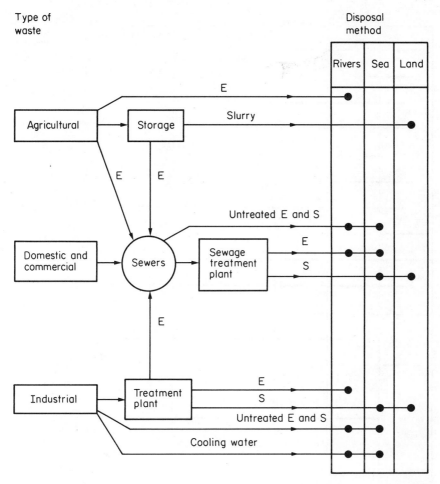

Figure 14.3. Outline of the disposal of waste solids and effluents: E, liquid effluent; S, sludge.

or without particles in suspension in it, which is discharged from premises used for carrying out any trade or industry, other than surface water and domestic sewage'. Trade premises in this definition includes all those used for industrial production and processing; and for agricultural, horticultural, scientific research and experimental purposes.

The largest volume of discharged waste is in the form of an effluent. This is a broad term that may be used to describe any solid, liquid, or gaseous product, in a treated or untreated condition, that is discharged from a process. In relation to water pollution, effluents are usually liquids that vary considerably in composition. For example, industrial effluents may contain water, organic solvents, oils, suspended solids, and dissolved chemical compounds. The chemical content may be organic or inorganic, and the effluent may vary in quality or strength

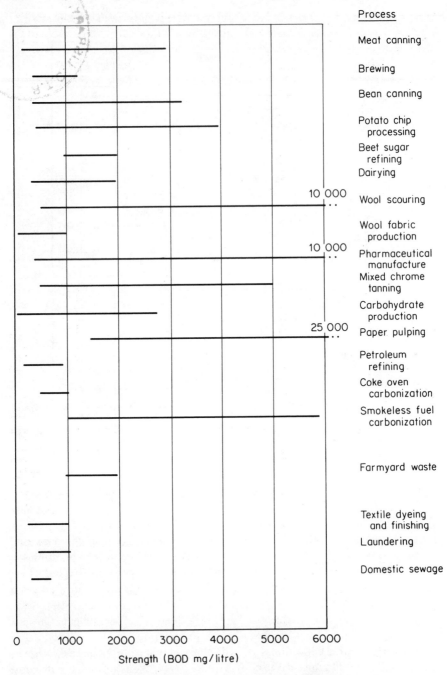

Process

Meat canning

Brewing

Bean canning

Potato chip
  processing

Beet sugar
  refining

Dairying

10 000 Wool scouring

Wool fabric
  production

10 000 Pharmaceutical
  manufacture

Mixed chrome
  tanning

Carbohydrate
  production

25 000 Paper pulping

Petroleum
  refining

Coke oven
  carbonization

Smokeless fuel
  carbonization

Farmyard waste

Textile dyeing
  and finishing

Laundering

Domestic sewage

Strength (BOD mg/litre)

0   1000   2000   3000   4000   5000   6000

Figure 14.4. The strength of some effluents.

from relatively harmless dirty water to highly toxic metallic and organic sludges, as shown in Figure 14.4. There are as many different types of effluents as there are industrial processes, so it is only feasible to consider effluents in general terms. They may be broadly categorized according to their origin as follows:

1. Sewage works effluent. This comes from domestic and commercial premises and often contains some industrial waste. It is normally treated before discharge into water courses, but there is some discharge of untreated sewage into the sea.

2. Industrial plant effluent. This originates from all manufacturing and processing industrial plants, and may be treated by the producer, or discharged into sewers and treated as sewage.

3. Agricultural premises effluent. This originates from dairy milking parlours, cattle housing premises, and silage clamps, and it is usually untreated.

4. Waste tip seepage. This effluent moves from domestic and industrial waste tips, mining and quarrying waste spoil heaps etc. into the ground, and eventually into water courses. It is usually untreated.

5. Run-off from land. This is untreated natural drainage of ground water into water courses.

6. Accidental spillage of chemicals during loading and transit. This occurs spasmodically and locally when the chemical is usually washed away into rivers, drains, and sewers.

7. Accidental leakage from industrial storage tanks, oil refineries, etc., sited beside rivers, estuaries, or the sea coast.

8. Other sources such as river and canal craft toilets, overloaded toilets on camp sites, and farm pesticide stores.

Details of the more important types of industrial effluents and their pollution strength or potential degree of pollution are shown in Figure 14.4. This shows that paper pulp and pharmaceutical manufacture, coal carbonization, wool scouring and tanning produce effluents with a very high BOD value; and others include silage, with up to 50 000, and some chemical effluents with up to 30 000 mg/l BOD. Each of these effluents contains a number of potentially polluting chemical substances, and some of these are listed in Table 16.2.

The agricultural industry produces considerable quantities of very potent organic effluent. The modern intensive housing techniques used for cattle, pigs, and poultry produce large amounts of manure slurry. This consists of excreta, urine, bedding materials, and pesticides derived from fodder and silage, as well as nitrates, and phosphates. Farmers have to dispose of this waste slurry, and the method used can create water pollution. At present there is no obligation upon the farmer to treat the effluent, but it must not be discharged directly into water courses and cause obvious pollution. The slurry is usually stored in tanks, and spread or sprayed on to arable and grass land at appropriate times of the year. If the slurry is applied in large quantities, or to heavy soils in wet conditions, there is land surface run-off and an increase in the ground water. Consequently surface drainage and underground leaching and seepage can cause organic pollution, and an increase in the concentration of nitrates and phosphates in water courses. This latter effect is increased where substantial quantities of nitrogenous and phosphatic fertilizers and pesticides are regularly applied to farm land and crops. Excess chemicals that remain unabsorbed by crops can leach out of the soil into ground water and eventually reach streams and rivers. Modern husbandry practices require the production of increasing quantities of grass silage for use as winter feed for cattle. Silage is made by mixing freshly cut grass with formaldehyde and acids in tower silos, and allowing bacteriological action to occur. Effluent is produced from the silos and it has a large organic content, so that the BOD value may be as high as 50 000 mg/1. The effluent should not be discharged into water courses and is usually run on to the land, but accidental leakages of silage cause a number of pollution incidents each year.

Marine waters are being increasingly polluted by crude oil. This is caused by the spillage or accidental discharge of oil effluents from various sources. Worldwide, there are probably more than 800 M tonnes of oil being transported by sea per annum, and inevitably there will be occasional accidents when ships collide or go aground. All oil tankers need to wash out their tanks between loadings, and some captains illegally discharge their wash effluents at sea causing localized pollution. The recommended procedure of the Clean Sea Code 1968 for tank washing is the 'load on top' system. This involves transfering tank washings into a slop tank, allowing the oil to float to the top, pumping out the water, and retaining the oil in the slop tank. If all tankers adopted this system there would be a significant fall in marine oil pollution. Apart from tankers, there are other accidental and intentional discharges of oil into surface waters. These occur in relatively small quantities from storage tanks, pipe lines, oil bunkering operations, oil refineries, and road transport accidents, and by private car owners. Some of this leaked or discarded oil effluent is discharged into sewers, or dumped on land, or into streams, and it can eventually drain into rivers and the sea. The development of off-shore drilling and oil well operations also creates spillages, and accidents can occur. The first North Sea oil well blow-out at the Ekofisk Field in 1978 was a recent example.

This brief survey of effluents shows the different sources, composition, and disposal methods in use. The specific chemical content of effluents is discussed in Section 16.3. It is important to consider the quantities of different effluents that are produced, because the more that has to be disposed of in the environment, the more likelihood there is of water pollution arising.

14.3 Quantities of Effluents

The total quantity of waste effluents that are discharged into all types of water in England and Wales is difficult to assess. Table 14.4 shows data obtained from several sources. The data relate to the years 1970 and 1973, and although the picture is incomplete, it enables a comparative study to be made. Inland and coastal waters received about 99 Mm^3 of discharged effluent each day. Of this total, 68% was cooling water which causes little pollution, but the remaining 32% was sewage and industrial effluent that is potentially highly polluting. Rivers received about twice as much effluent as the estuarine and coastal waters, and this

Tabe 14.4.   Effluent disposal in England and Wales

Type of effluent	To estuarine and coastal waters	To rivers	Total
Treated sewage	2.9	8.8	11.7
Untreated sewage	2.05	1.2	3.25
Industrial	4.89	11.0	15.89
Total	9.84	21.0	30.84
Cooling water	3.5	63.0	99.5

(column group heading: Quantity (Mm^3 per year))

was discharged from sewage works and industrial sites in about equal proportions (Figure 14.5). Figure 14.6 shows that 51% of industrial effluent was discharged directly into rivers and coastal waters, but a further 24% was discharged into sewers, together with 25% of domestic sewage. Therefore industry produced about 23 Mm^3 of effluent per day, and 32% of this was processed through sewage works, and 68% was dealt with by the industrial producers.

The above data do not show the extent of treatment of the effluent before discharge. In Table 14.4, 22% of the total sewage was described as untreated, which means that it received only cursory primary treatment. Almost twice as much untreated sewage was discharged into estuarine and coastal waters as into rivers. However, this does not account for all the pollution present in the rivers of England and Wales as shown in the national surveys. Therefore it seems possible that some of this pollution is produced by the discharge of untreated or insufficiently treated industrial effluent. This occurs, despite the control that should be exercised through the consent discharge system for industrial effluent.

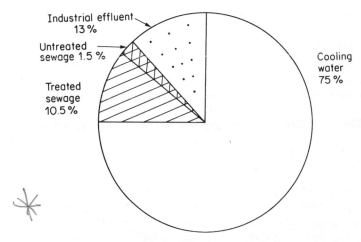

Figure 14.5. Effluent discharges to rivers in England and Wales.

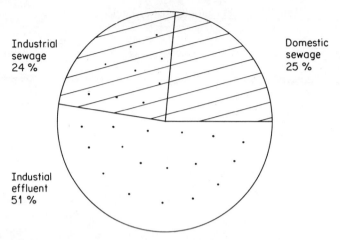

Figure 14.6. Effluent discharges to all river, estuarine, and coastal waters in England and Wales.

Some marine waters are used for the disposal of specific solid wastes. Along the north-east English coast, approximately 2.5 M tonnes of colliery waste is tipped into the sea from cliff tops, and 1.5 M tonnes are deposited on the foreshore. A further 1.5 M tonnes of coastal colliery waste is dumped at sea beyond the 5.6 km limit. North-east coastal waters also received 0.5 M tonnes of pulverized fuel ash from CEGB power stations in 1972. Therefore, marine waters between Blythe and Hartlepool are used to dispose of 6 M tonnes of solid waste annually. In the south-west about 1.5 M tonnes of china clay wastes are discharged into river estuaries and carried out into St Austell Bay. In GB, over 13 M tonnes of sand and gravel were dredged from the sea in 1970, and this was 10% of the total production for that year. Dredging operations only take place in restricted areas off the east coast, the southern North Sea, and in the Solent, Bristol Channel, and Liverpool Bay. Both the dumping of solid waste and dredging cause localized marine pollution. The water becomes turbid, and sludge is deposited on the sea bed causing a decrease of light intensity and unsuitable conditions for most marine organisms. In addition, material may be deposited on the sea shore which will severely affect the enjoyment and amenity of the coast. The Ministry of Agriculture, Fisheries and Food (MAFF) give consents for the discharge of waste into coastal waters. The consents and the quantities involved in 1971 were as shown in table 14.5. Note that the colliery waste figures exclude shore-tipped material, and the sewage sludge figures include disposals by Local Authorities, and are for wet weights.

There is environmental concern about specific effluents that contain radioactive pollutants. Some indication of the amount of this type of waste and its origin is shown by the data in Table 14.6. The total of 156 204 curies was only 32% of the permitted maximum agreed for radioactive liquid waste discharges. All discharges into the environment are controlled through authorizations granted by

Table 14.5. MAFF consents for discharges to coastal waters in England and Wales, 1971

Discharge area	Sewage sludge	Mixed industrial	Colliery	Mixed mineral	Fuel ash	Total
	Waste material to the nearest thousand tonnes					
Irish Sea	559	126*				685
North Sea	5133	293	1524	23	610	7583
English Channel		87			3	90
Bristol Channel	371	29				400
Total	6063	535	1524	23	613	8758

*Includes domestic waste.

the DOE, and they are monitored by the MAFF Fisheries Radiobiological Laboratory. Although the annual discharge of radioactive effluent is relatively low at present, the quantity being produced is increasing as more nuclear plants and installations are built, and fuel processing is expanded. The accumulation of radioactivity in food chains, and its effect upon the environment, needs to be very carefully researched and controlled in the future.

The total quantity of oil discharged into inland and marine waters is not known accurately. However, some indication is shown by the recording of spillage incidents and the estimated amount of oil involved (Figure 14.7). The number of incidents is increasing, according to the annual reports of the Advisory Committee on Oil Pollution of the Sea (ACOPS). There were 500 reported incidents in 1975, 595 in 1976, and 648 in 1977.

About 55% of oil pollution incidents occurred at sea, and 19% were in large ports such as London and Manchester, and the estuaries of the rivers Clyde, Forth, and Medway. The remaining 26% of incidents took place around the coasts of GB. Oil discharged at sea does not degrade and sink very rapidly, and so oil slicks are always a potential source of shore pollution. The size of these slicks

Table 14.6. Major discharges of liquid radioactive waste to estuarine and coastal waters in England and Wales, 1968–70

Source of waste	Authorized discharge (curies per year)	Mean discharge rate (curies per year)	Percentage
Fuel reprocessing plants	422 640	152 450	36
Reactor research and development	30 000	1 623	5.4
Nuclear power stations	18 565	1 151	6.2
Fuel fabricating plants	12 360	973	7.9
Naval establishments	71	6.6	9.2
Total	483 636	156 203.6	32.3

152

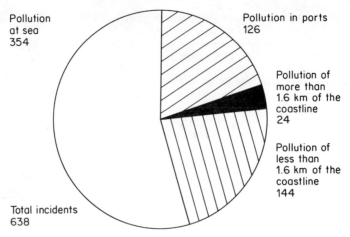

Figure 14.7. Oil pollution incidents around GB, 1977.

can be assessed by their effects, so that 4% of incidents caused pollution of over 1.6 km of the coastline, and 22% extended over less than this distance. Larger spillages caused by accidents to oil 'super' tankers are well known and reported. For example the Torrey Canyon released about 120 000 tonnes of oil in 1967, the Pacific Glory leaked about 49 000 tonnes in 1970, and the Amoco Cadiz released about 170 000 m^3 of oil in 1978. Marine oil wells can have accidents also, and the Santa Barbara oil well blow-out in 1969 released between 10 000 and 112 000 tonnes of crude oil near the California coast. Some indication of the total, annual oil pollution in the UK and the US is shown in Table 14.7.

Oil spilt on the high seas does not produce much apparent pollution, but the effect is serious when oil slicks move towards the coasts. Shore oil pollution affects all forms of aquatic life, it renders the shore unsuitable for seaside recreation and enjoyment, and it is very costly to clean up.

All the types of waste effluent so far described are directly discharged into water, under man's control. But some pollution is caused by the indirect entry of chemical substances into water courses, from waste deposited on land. Mining operations for coal and mineral ores produce large quantities of solid waste which is tipped on to spoil heaps, or as in-fill to quarries and disused mine shafts.

Table 14.7. Oil discharges into water

| | Tonnes per year | |
Source	UK 1972	US 1973
Land based discharges	780 000	3 700 000
Ship operational losses	735 000	750 000
Accidental discharges	196 000	1 380 000
Off-shore production	147 000	283 000
Total	1 858 000	6 113 000

Table 14.8.   Agricultural waste, 1971 and 1975

Type	Population in millions	Quantity of waste (Mm^3 per year)	BOD value of crude waste (mg/1)
Pigs	7	6.53	20 000
Cattle	12	29.56	40 000
Laying hens	41	13.6	55 000
Silage (1975)		7.73	60 000

The spoil becomes soaked with rain water and any pollutants present are leached out, enter the ground water, and eventually reach streams and rivers. Industrial wastes are also tipped as land-fill, and water pollution can occur in a similar manner, if the leachate is not properly drained away or contained by the geological strata around the tips. Agricultural husbandry wastes are not usually discharged directly into sewers or water courses. Animals reared and maintained in intensive housing units produce considerable quantities of manure and dairy husbandry effluent. For example, a 40 cow unit produces 2.7 m^3 of effluent per day, or one cow produces as much organic waste as 16 people. An estimate of the annual agricultural waste produced in England and Wales is given in Table 14.8.The disposal of an estimated 57 M m^3 of waste on to land presents considerable problems for farmers. Surface run-off and leachate from this land disposal can cause organic water pollution. Considerable quantities of fertilizers are used on British farms and an increasing proportion of these are reaching rivers through soil percolation and drainage. For example, the waters of the River Thames contained five times as much nitrate in 1977 as in 1948. The Water Research Centre, Stevenage, has estimated that the Great Ouse receives drainage water containing 60 kg of phosphorus, and up to 23 kg of nitrogen per hectare of land per year. Also the Centre for Agricultural Strategy Report 1978 stated that of the 180 000 tonnes of phosphate used annually, 33% is leached out of the soil.

All these different types of effluent and their pollutants may directly or indirectly enter water courses. It is the method of disposal, and the composition and quantity of the effluents, that determines whether they cause pollution. The effluents have to be disposed of within the environment, but there are treatments that can be carried out which will reduce the degree of pollution that may be created.

14.4 Effluent Treatment

Ideally no effluent should be released into the environment before it has received satisfactory treatment. The Water Authorities are attempting to reduce the amount of water pollution in GB by requiring improved methods of pre-discharge effluent treatment to be carried out. Effluent discharge to sewers, inland waters, tidal rivers, and coastal waters (i.e. the sea within 5.6 km of the shore), must be authorized by a disposal licence or consent, granted by a WA under section 31 (2) of the Control of Pollution Act 1974. The disposal licence includes conditions relating to the composition and quantity of the effluent to be discharged. It is the responsibility of the discharger to comply with the conditions of the disposal licence, and carry out any pre-treatment of sewage or industrial effluent that is required.

Treatment and Disposal of Sewage

Sewage is a turbid liquid, consisting of 99.9% water containing a complex of organic and inorganic matter, in the form of visible suspended solids, very finely divided suspended solids, colloidal particles, dissolved compounds, and microorganisms such as protozoa, bacteria, and viruses. The objectionable odour and colour is largely caused by the 66% organic matter present, and the anaerobic bacterial action that takes place within it. Organic matter is present as paper, faeces, urine, soap, detergents, fats, oils, greases, and food materials. The inorganic substances present include sand, clay, ammonia and ammonium salts derived from the decomposition of urine, metallic salts, nitrates, phosphates, etc. The precise composition of sewage varies according to its origin and its industrial effluent input, which may be up to 50% by volume.

Sewage treatment has been carried out in GB since the enactment of the River Pollution Prevention Act 1876, but the early methods were very rudimentary. Either the crude sewage, or the effluent after settlement of suspended matter, was allowed to percolate through soil, where natural soil bacteria acted upon the organic matter. The treatment was carried out at sewage farms, but was unsatisfactory because it needed large areas of land which quickly became obnoxious and soured, and there was a possibility of pollution of ground water and water courses. Alternatively, untreated or raw sewage was discharged into rivers, where dilution and natural bacterial action was relied upon to eliminate any pollutive substances. This practice still continues, especially in relation to effluent discharge into estuaries and coastal waters. The growth of population, the development of industry, and the establishment of large urban areas during the late nineteenth and early twentieth centuries caused the early treatment methods to become quite inadequate. Biological filtration was introduced about 1900, and gradually as new treatment plants were built for the larger towns, the modern sewage treatment system was developed. This is reasonably efficient provided that the through-put does not exceed the plant design capacity. However, the growth of technological developments in industry, and the use of an increasingly wide range of organic chemicals, has placed a considerable strain upon older sewage plants.

Modern sewage treatments basically aim to remove the floating and suspended solids, and provide biological treatment of the organic matter present. The resultant sewage effluent should be stable, have a low organic content, and be suitable for discharge into a nearby water course without causing pollution. Sewage treatment can be conveniently divided into three stages called preliminary, primary, and secondary, with some plants providing a final tertiary

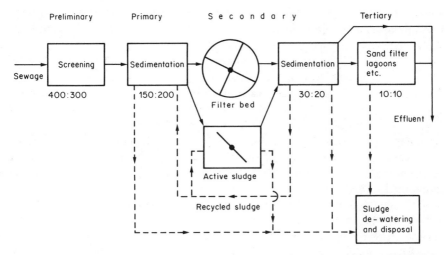

Figure 14.8. Flow diagram of a sewage treatment plant. The numbers refer to the SS:BOD strength of the sewage (see text). The full lines indicate sewage flow, and the dashed lines sludge flow.

stage to produce a higher quality effluent (Figure 14.8). The composition of sewage can be expressed in terms of the quantity of organic matter present and its biological oxygen demand or BOD (Section 15.3). Raw sewage entering the plant shown in Figure 14.8 may contain a total organic content of 1000 mg/l, with a suspended solid (SS) value of 400 mg/l, and a BOD of 300 mg/l. Hence its strength is described as an SS : BOD of 400 : 300. At the end of the primary stage the sewage has a value of 150 : 200, and when the secondary stage is completed the value of the effluent is reduced to 30 : 20. This is the normally accepted standard that was originally proposed by the Royal Commission on Sewage Disposal Eighth Report in 1912, and has been used ever since. If the tertiary treatment stage is carried out, a purer effluent of 10 : 10 strength can result. The preliminary treatment removes large suspended debris and solid particles by screening. In the primary stage, sedimentation takes place which results in the removal of some 55% of settleable solids as sludge, and a 35% reduction of the BOD value. The sewage now contains suspended colloids, finely divided solids, and dissolved solutes.

There are two alternative methods of secondary treatment in use, but both depend upon biological action of organic matter. In the older type of treatment the primary settlement tank effluent passes through so-called filter beds of stone or clinker. It is not a filtering process because the suspended organic matter percolates through the bed, and it is not separated from the liquid effluent. The solid material in the percolating or trickle filter beds provides a large surface area for the growth of a biological population. This is present as a slimy film containing bacteria, fungi, protozoa, insect larvae, and small worms. The more modern treatment is the activated sludge method, where the primary effluent is agitated and aerated in large tanks, in the presence of a flocculent suspension of activated sludge containing micro-organisms. This method is favoured for treating larger flows and requires less land than a series of filter beds. In both these methods of biological treatment, micro-organisms utilize the organic matter and break it down into non-pollutive inorganic compounds. The process is described in detail in Section 15.4. The treatment has been called self-purification, and it takes place naturally in streams and rivers. However, in sewage works, optimum conditions are provided to enable the biodegradation of organic matter to take place more rapidly than in natural waters. Following the secondary biological treatment there is a further sedimentation phase to allow more settlement of suspended solids. In a reasonably efficient plant, the effluent emerging from the secondary stage will have a 30:20 standard, and can be discharged into water courses.

If a higher standard effluent is required, or it is necessary to remove specific pollutants, then tertiary treatment can be carried out. The type of treatment varies according to the operative requirements, but it can involve flocculation, sedimentation, slow sand filtration, microstraining, pebble bed clarification, surface irrigation, and retention in lagoons. Tertiary treatment has its limitations, for example it cannot completely compensate for poor secondary treatment, and it does not remove all suspended solids, organic matter and dissolved solutes. Consequently the effluent cannot be used as drinking water without the normal chemical treatment given to natural waters.

Primary and secondary treatment aims to produce an effluent that can be discharged into water courses without causing pollution. The treatment does not completely remove all potentially polluting material, especially metals, nitrates, phosphates, ammonium compounds, and bacteria. Also all sewage works do not use the same treatment processes or operate to the same level of efficiency. A 1978 Water Authority review showed that only 33% of the Thames works, 50% of the E. Anglian works, and 23% of the North West works produced completely satisfactory sewage effluents with the recommended 30:20 BOD value. There are several explanations given to explain this unsatisfactory situation. Many sewage works are old and obsolete, and their designed treatment capacity is far below present day demands. During periods of excessive treatment demand, these works have no alternative but to by-pass some processes, and they discharge only partially treated effluent. This situation also occurs when there is heavy rain, and the storm water flow of surface and sub-soil water overloads the treatment capacity. Sewage works that were constructed in the first half of this century were not designed to process the type of effluent they now receive from modern industrial plants. One instance of this occurred in the late 1940's, when the use of organic synthetic detergents began to replace soaps. These so-called 'hard' detergents were not readily biodegradable, and they were only partially decomposed during the secondary treatment stage. The detergents also contained additives, polyphosphates, and foaming agents, which resulted in sewage effluents being discharged containing boron and phosphate, and masses of foam appeared on rivers and streams. The problem has now been eased by manufacturers agreeing to use 'soft' or biodegradable organic detergent bases, but phosphates are still present in the discharged effluents. The increasing industrial use of toxic organic compounds, that may pass untreated through sewage works, is another cause of water pollution. Untreated and treated sewage effluent can contain viruses responsible for water-borne human infections. This is being investigated in the UK and the US, because of the increasing incidence of some viral illnesses, such as infectious hepatitis, gastroenteritis, and salmonella infections. At present there is little testing of sewage effluent for the presence of viruses, and no quality standards exist.

Sewage treatment produces large quantities of waste sludge, which results from the primary and secondary sedimentation processes and any tertiary treatment that is carried out. Small amounts of sludge are recycled back into the treatment process to maintain the micro-organism population necessary for sewage biodegradation, but the remainder must be disposed of within the environment (Figure 14.8). It is estimated that about 1.4 M tonnes of dried sewage solids are produced in the UK. Sludge disposal is a major problem for all types of sewage works, and in 1978 cost £35 M or 40% of the total treatment cost. Sludge is treated before disposal to reduce the volume, and to ensure that there is no objectionable odour, or severe contamination with organisms or toxic substances liable to cause pollution. The treatment usually consists of digestion for 25 to 30 days in closed tanks heated to about 30–35°C. This allows rapid microbiological action, or 'alkaline fermentation' to take place under anaerobic conditions. Sludge organic acids are produced, and these then decompose to produce carbon dioxide and methane. This heated digestion process removes odour, kills most pathogenic bacteria, and reduces the sludge solid content by up to 40%. The methane gas yield is sufficient for utilization as a fuel for vehicles, site gas engines, and heating boilers used for sludge digestion, drying, or incineration. As a result of the anaerobic digestion process the sludge volume is reduced, but it may still contain up to 50% water. So it is usually dried or dewatered in drying beds or lagoons, or by means of centrifuges, vacuum filters, filter presses, or rotary heat driers. At the end of this treatment, the dried sludge must be disposed of. Four methods are in common use, namely spreading on agricultural land, landfill in old quarries and mineshafts, dumping into the sea, or processing for use as a fertilizer or soil conditioner. In 1978, about 20% of sewage sludge was dumped on land, 40% dumped at sea, and 40% was used by farmers as a basis of nitrogen and phosphate fertilizers, equivalent to £15 M per year. The returning of sewage waste to the land may appear to be environmentally advantageous, but compared to manufactured NPK fertilizer, digested sludge may be deficient of up to 66% of nitrogen, 85% of phosphate, and 90% of potash. The use of sewage sludge as a fertilizer is not favoured by some farmers, because of possible contamination, and the fact that its nutrient content is very variable. However this long established practice is still widely used. On some farms, solid sludge may be ploughed in once a year before the next arable crop is sown. Sludge liquor may be sprayed on to grassland several times during the growing season. This practice does help to condition the soil by increasing the humus content and so maintain good drainage and fertility. However if the sewage treatment has not been satisfactory, there is a danger that heavy metals, organic and other pollutants may affect plants and animals, and contaminate ground water. Similarly, landfill using sewage sludge can cause water pollution if there is seepage of pollutants from the dumping site into ground water. The MAFF and their advisory service (ADAS) issue advice to farmers so as to try to safeguard the possibility of land toxicity in respect of zinc, copper, lead, and chromium. Continuous application of contaminated sewage sludge and effluent can cause toxic concentrations of these metals to occur in the soil and pollution of water courses. Recent advice has proposed that grassland should not be used for grazing within 3 months of spraying with treated sewage, and 6 months if untreated material is used. Also no crops should be eaten until about a year has elapsed after the use of treated sewage on the land. Large cities such as London, Manchester, Liverpool, Bristol, and Glasgow dispose of their sludge by dumping it at sea, usually at a location at least 24 km from the coast. Unlike the practice in land disposal, the sludge is raw or only partially treated, but it may be dewatered to reduce bulk and transport costs. Theoretically, the dilution of sewage sludge in the sea should prevent pollution, but this is questioned by some authorities. There is a potential danger if persistent toxins build up in the dumping areas or elsewhere, and they eventually enter marine food chains.

Modification of sewage treatment processes, and more specialized tertiary treatment plants will be required in the future to overcome present problems. Considerable quantities of virtually raw sewage are still being released into surface waters and the sea, and there is no overall legal requirement controlling this practice. Probably two-thirds of coastal towns carry out no systematic sewage treatment, and thousands of private houses, hotels, caravan and camping sites located in coastal areas, do not discharge their waste into the public sewerage system.

Treatment and Disposal of Industrial Effluents

Industrial wastes of very varied composition are treated and disposed of in a number of ways. Many small industrial companies discharge untreated wastes into the public sewerage system, and these are processed with domestic sewage at the Water Authority's works. Larger industrial firms are usually responsible for carrying out their own waste processing and disposal. Industries such as the CEGB, the National Coal Board, and quarrying, remove suspended solids by sedimentation, and discharge the effluent directly into water courses. Other industries, especially those producing very toxic wastes, may transport the wastes away to land disposal sites, without treatment. Some large industrial concerns process their waste in on-site treatment plants, dump the solid sludge, and discharge the liquid effluent into the public sewers. Both large and small companies may be required to carry out some type of treatment on site, prior to discharging effluent into the sewerage system. Whatever the treatment system, the resulting liquid waste effluent from any industrial site is discharged into water courses under the consent procedure of the local Water Authority. This usually requires that any discharged effluent should be free of floating solid matter and oils, free from heavy suspended mineral and organic matter, and not contain significant amounts of biologically inert or toxic substances which will affect the activity of natural aquatic micro-organisms.

In general terms, industrial effluents are treated to remove suspended solids by screening, precipitation or flocculation, and sedimentation, followed by biological and/or chemical treatment as appropiate. Effluents with a high biodegradable content can be processed in a similar manner to that used for sewage. Other effluents may be very acid or alkaline, or contain biologically inert substances, or toxins. These require chemical, rather than biological treatment, and some of the processes carried out are as follows:

(1) coagulation and precipitation for the removal of solutes, suspensions and emulsions;
(2) natural or air-floatation to separate oils and greases;
(3) addition of alkalis to raise the pH of acid wastes;
(4) distillation to separate organic and other solvents;
(5) steam or inert gas stripping to remove low-boiling-point wastes;
(6) oxidation, for example the conversion of sulphides to sulphates;
(7) electro-dialysis, or ion exchange, or active-carbon adsorption to remove highly toxic substances.

A number of industrial firms use some of the above processes to extract or recover useful chemical substances during the waste treatment, for example fats, oils, metals and metallic compounds, carbohydrates, and solvents. The recovered substances can either be recycled within the firm's industrial production, or offered to other companies for use as basic or intermediate chemicals. The exchange of recovered chemicals is facilitated through the Government Warren Spring Laboratory Waste Recovery Service, and private industry chemical recovery firms.

The overall aim of industrial waste treatment is to separate the suspended solids and dissolved substances from the liquid fraction. The latter is usually discharged into sewers or water courses, according to its composition. The separated solids are usually in the form of a sludge, and the disposal of this presents as many problems as for sewage sludge. Limited surveys of a sample of industrial companies in 1970 showed that 72–88% of waste was tipped on land, 3–16% was dumped at sea, and 0.5–1.1% was incinerated. Approximately 60% of the waste tipped on land was in the form of a sludge, and contained a very wide range of chemical substances including toxic compounds. Indiscriminate land tipping can cause water pollution, where contaminating substances leach out of the waste into the ground water, and subsequently into streams and rivers. Indeed in the late 1960's and early 1970's, there were numerous reported incidents of unowned 'leaking' tips causing water pollution. As a result, the Deposit of Poisonous Waste Act came into force in 1972, requiring the licensing of such tips. Now many industrial companies employ the services of waste disposal contractors to transport the waste to their own approved sites for properly controlled tipping. Water pollution derived from industrial waste disposal still occurs, despite the legislation that exists to control land tipping and effluent discharge into water. This is mainly because the Water Authorities are unable to monitor continously all licensed tips and discharges, and many small industrial concerns do not adopt a sufficiently responsible attitude towards pollution prevention. The Water Authorities are attempting to tighten the conditions of discharge consents, and require more and improved on-site treatment of waste to be undertaken by industry. This will take time to implement, it will require capital finance, and involves the question of who pays for pollution abatement.

CHAPTER 15

Biological Effects of Water Pollution

A. S. Wisdom gave a legal definition of pollution in 1956, as 'the addition of something to water which changes its natural qualities, so that the riparian owner does not get the natural water of the stream transmitted to him'. So to understand the biological effects of pollution it is necessary to appreciate the natural conditions that exist in surface waters. In fact, when water recovers from pollution it returns to the natural balanced state, and the recovery is brought about by natural biological processes.

15.1 Natural Waters

Natural waters cannot be easily defined, but they can be described in terms of the physical, chemical, and biological conditions present. The physical condition should consist of clean water with an ambient temperature, and freedom from most suspended solids, colouration, surface scum or foam, obnoxious odour, and taste. The chemical condition should ensure that there is adequate oxygen, a correct balance of dissolved chemical nutrients to support life, and an absence of excessive organic matter, and toxic substances. Biologically, most natural waters contain a range of micro-organisms, plants, and animals that exist in a balanced ecological state. All ponds, lakes, streams, and rivers are ecosystems, where the biological population exists within the physico-chemical environment. Each ecosystem has its own organization, comprising biological nutritional cycles for the continued survival of the species constituting the food web within the system. Any change that takes place within the ecosystem caused by the introduction of external material may be regarded as pollutive. Natural pollution can occur, perhaps caused by leaf fall or dead animals, whereby toxins are introduced into the water, and these change the ecological balance and the physico-chemical conditions. Other forms of pollution are induced by human activities, and the effects may range from slight to heavy. Slight or natural pollution produces no visible abnormal effect upon the water, and is said to be absorbed. This means that the ecosystem quickly readjusts, and returns to its natural state with little or no effect upon the water or the life forms. Slight pollution occurs in many natural waters and is not an environmental problem.

15.2 Dissolved Oxygen Content of Water

Oxygen is vital for the respiration of nearly all biological life, and it is essential that water should be well aerated for the continued survival of aquatic life. Aeration involves atmospheric oxygen diffusing, or passing from the air, through the water surface into ponds, streams, rivers, canals, and the sea. Water in a turbulent stream or passing over a waterfall or weir has a large surface area, and so it takes up oxygen rapidly and is well aerated. In contrast, a slow moving river in summer has a low oxygen content. Oxygen is not very soluble in water but it does exist in solution as dissolved oxygen. The amount available in any volume of water is expressed as parts per million by weight (ppm), or milligrams per litre (mg/l), or grams per cubic metre (g/m^3). The maximum amount of dissolved oxygen that can be held in solution in water is expressed as the saturation percentage, and this varies with the temperature and chemical composition of the water. For example 100% saturation of fresh water is equivalent to 14 mg/l of oxygen at 0°C, or 9 mg/l at 20°C, or 7.6 mg/l at 30°C. Sea water, with a comparatively higher salt content, has a lower oxygen saturation percentage than fresh water at the same temperature, for example 7 mg/l at 20°C. Supersaturation of up to 200% can occur in a turbulent stream with bright sunshine and a high ambient temperature, and where a high rate of photosynthesis produces additional aeration of the water.

The oxygen saturation percentage in water is a vital factor, which affects

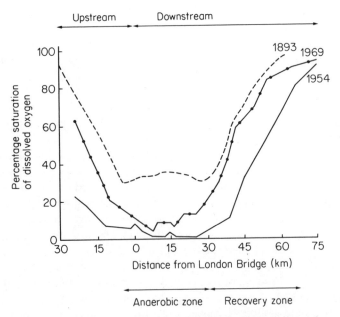

Figure 15.1. Oxygen sag curves for the River Thames (from the Royal Commission on Environmental Pollution *First Report* 1971).

support for the aquatic ecosystem and the resultant effects of pollution. When organic waste is discharged into a river, some of it is oxidized by the natural micro-organisms present, and this creates a biological oxygen demand or BOD. Consequently the dissolved oxygen level and the saturation percentage falls, and an oxygen deficit is created. If the BOD and deoxidation rate exceeds the reoxidation rate from the atmosphere, then the oxygen deficit persists. These are the conditions of water pollution and disruption of the aquatic ecosystem. If no more organic matter enters the water, the existing pollution material is gradually completely oxidized, and so reoxidation takes place and the river is said to recover from the pollution. This process can be illustrated by an oxygen sag curve as shown in Figure 15.1. The figure shows the sag in the oxygen saturation percentage caused by severe pollution in the River Thames, and this extended over a distance of about 80 km in 1954. The degree of pollution became progressively worse during the first 50 years of this century. Indeed a survey in 1957–58 showed that there were no living fish present in the river between Richmond (24 km upstream from London Bridge), and Gravesend (40 km downstream). During the late 1950s, the Port of London Authority launched a programme to improve the water quality, and the results of this are shown by the 1969 curve. In 1967–68 it was found that 42 species of fish were now present in the polluted zone, and migratory forms were able to pass through it and survive. By 1977, 91 fish species were present.

15.3 Oxygen Demand Tests

The Royal Commission on Sewage Disposal, 10th Report in 1912, first adopted the well known biological oxygen demand (BOD) test as a measure of the polluting organic matter present in a sample of water. BOD can be defined as the amount of dissolved oxygen consumed by chemical and microbiological action when a sample of water is incubated for 5 days at 20°C in the dark. BOD is expressed as mg/l or ppm of oxygen taken up by the sample. When the test is carried out, the water sample is suitably diluted with aerated water and divided into two portions. The dissolved oxygen content of the first portion is estimated at once, and the second portion is incubated for 5 days, which is the time required for the oxygen consumption to reduce to a minimum under natural conditions. By comparing the dissolved oxygen content of the two portions, the BOD value is determined. The BOD test has the disadvantages of requiring 5 days to carry out, the oxidation of the organic matter is incomplete, and the microbiological action may be affected by toxic pollutants present. Therefore, although the test has been used over many years for sewage effluent, some people consider it to be unreliable, especially for industrial effluent. The permanganate value (PV), or chemical oxygen demand (COD) tests are in use as alternatives to the BOD test. The PV test determines the amount of oxygen taken up from a solution of acidified potassium permanganate, by the contents of a water sample at 27°C. This test gives consistent results, but only 30–50% oxidation of the organic and inorganic compounds present is completed. The COD test measures the amount

164

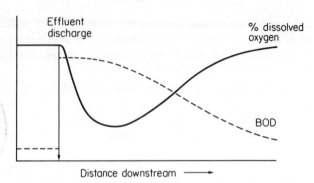

Figure 15.2. Oxygen saturation percentage and biological oxygen demand (BOD) downstream of an effluent discharge.

of oxygen absorbed from a solution of boiling acidic potassium dichromate in 2 hours, by the contents of a water sample. This test involves the oxidation of all organic matter in the water, and so it may over-estimate the amount of biodegradable matter present. A comparative study of the three tests shows that on average for domestic sewage, the ratio between the BOD and PV tests is 3.1 : 4.1, and for BOD : COD is between 0.2 : 1 and 0.5 : 1. Most of the published data on pollution is expressed in terms of the BOD level (see Figures 14.2 and 14.4). The relationship between oxygen saturation percentage and BOD can be shown graphically, by plotting dissolved oxygen against time. For an organic effluent, time can be proportionally represented by the distance downstream from the point of discharge. Figure 15.2 shows the dissolved oxygen and BOD levels before an effluent was discharged into the water, and the subsequent effects. There is a rapid decrease of dissolved oxygen near the pollution outfall, followed by a gradual increase with increasing time after discharge, to reach the original dissolved oxygen level of the unpolluted water.

15.4 Biodegradation

Organic matter is described as biodegradable if it is readily decomposed by the action of micro-organisms. These exist as a mixed species population of bacteria, fungi, and protozoa; and in heavily polluted waters they are visible as white, yellow, pink, or brown slimy growths that are often called 'sewage fungus'. The micro-organisms utilize pollutive organic compounds for their growth and nutrition and produce simpler products, thereby reducing the amount of pollution. They have two types of metabolism depending upon whether they require dissolved oxygen or not. Aerobic micro-organisms use oxygen to carry out oxidative reactions such as the following:

1. Carbohydrates, phenols, etc. are converted to carbon dioxide and water.
2. Organic nitrogen compounds are converted to carbon dioxide, water, amines, and ammonia.

3. Organic sulphur compounds are converted to sulphides.

4. Organic phosphorus compounds are converted to phosphates.

Some first formed products are further acted upon, for example the bacteria of the nitrogen cycle convert ammonia to nitrite and nitrate, and sulphur bacteria change sulphide to sulphate. All these reactions require oxygen, and so biodegradation rapidly decreases the dissolved oxygen content, and creates a BOD in polluted water. Heavily polluted waters have little or no dissolved oxygen present. Under these conditions only anaerobic micro-organisms can exist, and as a result of their activities a different type of biodegradation takes place, for example:

5. Carbohydrates are converted to methane.

6. Organic sulphur compounds and sulphates are converted to sulphides.

7. Organic phosphorus compounds are converted to phosphine.

8. Organic nitrogen compounds are converted to nitrate, and then ammonia.

Ponds, streams, and rivers that are heavily polluted often have an obnoxious smell caused by hydrogen sulphide, and may show a black deposit of iron (II) sulphide and heavy growths of sewage fungus.

The type of biodegradation that occurs is also affected by the content of pollutive discharges. For example, if the effluent contains biological toxins such as heavy metals, cyanides, and sulphides, the aerobic micro-organisms will be killed, even if there is sufficient dissolved oxygen in the water. A surface film of oil on water restricts reoxygenation, and together with bacterial degradation of the oil hydrocarbons causes the dissolved oxygen content to fall quickly. Hard detergents are resistant to secondary sewage treatment, and frequently produce a layer of surface foam when the effluent is discharged into rivers. This foam also restricts surface reoxygenation, and may cause anaerobic conditions to arise.

The process of biodegradation occurs continuously in all natural waters, and it plays an essential part in the maintenance of balanced natural cycles within an aquatic ecosystem. When pollutive waste enters a river, the balanced ecological conditions are changed. Downstream from the discharge point the rate of biodegradation increases, causing a fall in the dissolved oxygen content of the water, and an oxygen deficit as shown in Figure 15.2. The extent of these changes depends largely upon the total amount of organic material in the river, and three examples will illustrate this.

1. If the quantity is small, the oxygen deficit is quickly eliminated by surface reoxygenation, and the water appears to absorb the pollution. In this instance there is only slight pollution, and a temporary change in the ecosystem over a short distance of the river downstream of the discharge that is quickly readjusted.

2. Larger organic discharges, perhaps adding to upstream pollution already present, cause greater ecological changes. Downstream of the discharge a large oxygen deficit is quickly created and the ecosystem is severely disrupted. The changes that occur in species populations are complex and variable, but the general effects for a class 3 river with moderately severe pollution are shown in Figure 15.3.

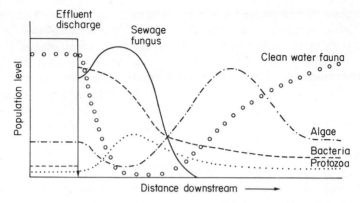

Figure 15.3. Effects of organic pollution on biological life in a river.

Near the point of the effluent discharge, the bacterial population rapidly increases and there is active growth of sewage fungus. Protozoa species that feed on bacteria increase in numbers, but the oxygen deficit causes a decrease of algae and fauna. Fish are particularly sensitive to dissolved oxygen concentrations and so they are often eliminated, and only a few invertebrate species can exist in low oxygen concentrations, for example sludge worms (*Tubificidae*) and blood worms (*Chironomus*). At the lowest point of the oxygen sag curve there is maximum biodegradation, and little plant and animal life compared to the unpolluted stretches of a river. Further downstream, there is a gradual recovery from the effects of the pollution. The biodegradation bacteria and sewage fungi populations decrease as their organic source of nutrition declines. The biodegradable process releases inorganic salts as shown in Figure 15.4, and the increase of nitrates and phosphates results in increased algal growth for a short distance along the river. Reoxygenation of the water provides improved aerobic conditions that allow the normal flora and fauna to slowly develop or reappear.

   3. The third example is a very severely polluted class 4 river. Here the rate of

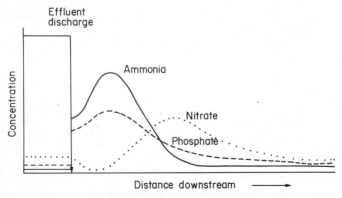

Figure 15.4. Effects of organic pollution on nutrients in a river.

biodegradation is high enough to cause depletion of all the dissolved oxygen, and the establishment of anaerobic conditions over part of the river length. The ecological changes outlined above follow a similar sequence, but the effects are more severe. There is complete elimination of all aerobic bacteria, algae, floating and fixed plants, fish, and most of the invertebrate animals in the anaerobic zone. Recovery from this pollution takes place more slowly, and extends over a greater length of river which may be as far as 65 km downstream from the source of pollution.

The effects of water pollution have been described in general terms, but there is considerable variation in different streams and rivers and under different weather conditions. Probably the most important factors are the quantity and composition of the organic matter discharged, and the volume of water available in the water course. If an effluent is untreated or only partially treated before discharge, it causes more pollution than a treated one. Secondary sewage treatment is a controlled biodegradation process, and sludge treatment is similar to anaerobic biodegradation. The maximum concentration of an effluent is at the point of its discharge into a stream or river. Downstream from this point the concentration is gradually reduced, as the pollutants are being biodegraded and diluted or dispersed by the water flow. In seasonal periods of low rainfall or drought, the water volume and flow rate are reduced. Therefore the dilution effect is less, and there is an increased pollution effect under these conditions. This also occurs if there is a large abstraction of water supplies from a river, and no adjustment of the water volume by the use of a compensating reservoir or other supply. Alternatively, when heavy rainfall takes place there is a greater dilution of the effluent after discharge, and less pollution is apparent. However, if a sewage treatment plant becomes overloaded by storm water, the discharged effluent may be untreated or partially treated, and so the river pollution is increased. These interacting factors also explain why upland stretches of rivers are relatively little polluted, because there is a high rate of flow, good dilution, and rapid atmospheric reoxygenation of the water. Lowland river stretches with a more sluggish flow are often much more polluted, especially if they receive one or more effluent discharges, and there is water abstraction (Figure 13.6). The resultant pollution that occurs in any one water course, and its recovery capability, are largely controlled by the interacting physical and biochemical factors operating within the ecosystem.

CHAPTER 16

Pollutants and their Effects

The effects of water pollution may be considered under six headings as follows:
  (1) physical effects, such as suspended particle solids that cause water turbidity, cooling water that raises water temperature, and oily surface films that restrict the reoxygenation of water;
  (2) oxidation effects caused by bacterial action or chemical oxidation of inorganic and organic substances, both of which significantly reduce the dissolved oxygen content of water;
  (3) toxic chemical effects caused by a range of substances that cause immediate or cumulative physiological changes in plants, animals, and humans;
  (4) chemical nutrient effects resulting from high concentrations of nitrates and phosphates;
  (5) pathogenic effects caused by micro-organisms, where bacteria and viruses are present in sufficient numbers to cause a health hazard;
  (6) radionuclide effects, caused by the accumulation of radioactive substances in food organisms, which produce human body changes.

16.1 Physical Effects

Various types of suspended solids are discharged into fresh and coastal waters. Solids may be inert mineral wastes such as china clay or mining spoil, or insoluble finely divided organic solids. The latter usually undergo slow biodegradation and cause a reduction of the dissolved oxygen in water. Inert solids are of varying particle size and density, and they either settle out, or remain suspended according to their properties and the turbulence of the water. Settled particles may slowly accumulate on vegetation foliage, and produce a deposit on the river bed. The effect of settlement layers is to reduce the solar energy absorption by plants and so lower the rate of photosynthesis, and to produce low oxygen conditions on the river bed. This can prevent the development of salmon and trout eggs, and preclude the survival of bottom living invertebrate animals. Small suspended particles make water turbid, and this reduces light penetration, reduces photosynthesis, and restricts plant growth. Turbidity also reduces visibility in the water and limits the food gathering capacity of many animals. Fish and some invertebrates have their respiratory efficiency reduced because the gill surfaces become clogged with suspended matter. All these physical effects cause a disturbance of the balanced ecosystem. Some animal species do not survive, others are reduced in numbers, and so food chains and nets are affected. The importance of the effects of suspended particles is shown by the fact that Water Authority discharge consents specify a permitted quantity of suspended solids, for example 30 g/m^3.

Cooling water from CEGB power stations is usually discharged directly into rivers, and this can cause a rise in water temperature and bring about thermal pollution. The biological effects of thermal pollution depend upon how much the temperature is raised. This is because the metabolic rate of physiological processes is speeded up by heat, for example for every 10°C rise in temperature the rate may be doubled. As each species has its own metabolic rate, most aquatic animals can only exist within a specific temperature range. For example trout are killed by a temperature of over 25°C and their eggs will not develop in water above 14°C, but carp can withstand temperatures of up to 35°C. One effect produced is that coarse fish such as carp, pike, perch, bream, and roach may only be found in rivers with thermal pollution present. Most fresh water fauna populations decline with rising temperatures, and few species can exist in a water temperature of over 40°C. Plants are less sensitive to temperature changes, but higher temperatures increase the rate of their physiological processes and speed up growth. Above about 30°C green algae tend to be less numerous, but there is an increased growth of blue-green algae and sewage fungus. This can eventually result in plant death and decomposition causing water stagnation. The overall effect of heated cooling water discharges depends upon the volume and temperature of the discharge, and the rate of flow and degree of pollution of the receiving river. The pollution effect will be greater in summer with an air temperature of over 21°C, and where there is a reduced water volume flowing in a sluggish river. The resulting rise in water temperature will lower the oxygen saturation percentage and speed up the biodegradation of pollutant organic matter. Both these effects will result in a sharp increase in the oxygen sag, or deficit, in the water. A rise in temperature also increases the toxicity of some chemical pollutants. The effect of thermal pollution on the ecosystem will obviously vary according to the interaction of both chemical and physical factors. There has been legislation to control thermal pollution since 1919. The temperature of cooling water discharges entering fresh water should not exceed 25°C, and not be over 20°C for sea water.

Waste oil, fats, and grease can enter rivers and estuaries from several sources. They are present in industrial effluents, and in addition can enter water through accidental spillage from oil refineries and terminals, oil storage tanks, road and rail tankers, and canal transport. Used oil is discharged into sewers through the irresponsible actions of private motorists and small garages. Oil and allied petrochemicals form a thin film on the water surface which prevents the exchange of oxygen with the atmosphere. This causes a reduction of the water oxygen saturation, and deoxygenation pollution effects arise. Spillage from oil tanker accidents and the illicit discharge of tank washings causes marine pollution and shore contamination. Oil forms large floating areas or slicks that remain at sea and eventually disperse and sink, or may be driven in shore through the action of wind and tides. At sea, oil slicks are responsible for the deaths of many birds such as scoters, auks, guillemots, puffins, razorbills, and diving ducks. These birds live on the water surface and dive to obtain their food. Oil coats the feathers externally, and this reduces their thermal insulation and resistance to cold, and prevents flight in search of food. The birds ingest oil when they dive to feed, and also when they attempt to preen their feathers to remove the oil. When oil is taken into the body it irritates the digestive system and produces toxic effects. The overall effect is to produce a high mortality, for example the Torrey Canyon incident in 1967 was probably responsible for the deaths of up to 100 000 birds. Very few birds that are badly contaminated appear to recover even after de-oiling and hand feeding. In-shore oil is deposited on rocks and sand, and this prevents the beaches being used for recreation and enjoyment by the public, and affects marine life. Shore animals such as crabs, shrimps, mussels, winkles, limpets, and barnacles ingest toxic hydrocarbons, they are unable to feed, and the respiratory system becomes clogged and ineffective. A badly oiled shore can be largely denuded of animal life, and seaweeds are also affected.

The specific effect and its duration is variable. It appears to depend mainly upon the composition and quantity of the oil, and the type of cleaning up operations that are carried out. In the Torrey Canyon incident, large quantities of solvent-based detergents were used for the first time in an attempt to quickly disperse the oil. It is now known that these detergents were highly toxic, and in some areas of the Cornish coast the normal balanced ecosystem was not restored for 10 years. However in other areas, where no detergents were used, the ecosystem recovered over a period of 6 months to 2 years. Studies have shown that if the oil contains light aromatic fractions these quickly evaporate, as in the Amoco Cadiz incident off the Brittany coast in 1978, where 30% of the spilled oil evaporated. The heavier oil fractions remain on the shore and are broken up by physical forces such as wave and tide action, UV light and evaporation, and also degradation by bacterial action. The best method to deal with oil pollution has not yet been determined. Newer water-based detergents have been developed that are claimed to be less toxic, but any organic detergent will be toxic to some extent. An alternative method is to remove the surface oil before it reaches the shore, but the present techniques only seem able to remove up to 10%. Some biologists advocate that no detergents should be used, and point out that the natural recovery period is less than when chemicals are used, and the total effect on marine life is less toxic. Whatever method of shore cleaning is used in the future, the necessity for it will be reduced, if there are strenuous efforts made to lower the incidence of oil spillage at sea.

16.2 Oxidation Effects

There are two main types of oxidation, brought about by the action of bacteria upon organic pollutants, or through chemical oxidation of other pollutants present in industrial wastes. Both types of oxidation involve the use of dissolved oxygen, and so produce an increased BOD and an oxygen deficit in water courses. Examples of bacteriological oxidation are the conversion of sulphide to sulphate in the sulphur cycle (Figure 2.5), and ammonia to nitrite and then to nitrate in the nitrogen cycle (Figure 2.2). Some chemical oxidation takes place in water where there are no suitable bacteria or insufficient numbers present. Another example occurs where drainage water from mines and spoil heaps enters streams and rivers. The drainage water often contains iron (II) sulphate and iron hydrogen carbonate. These iron salts are oxidized to iron (III) hydroxide, or ochre, which is deposited as rusty red gelatinous masses. These deposits are often associated with filamentous bacteria, and if present in large quantities are toxic to biological life. However most oxidation effects are caused by bacterial action upon organic waste discharged in sewage and industrial effluents, and this has been described under biodegradation.

16.3 Chemical Toxic Effects

Some inorganic and organic chemical substances are toxic or poisonous to plants, animals, and humans. A toxin may be described as any chemical substance that is capable of causing injury, or impairing, or killing any living organism. Toxins are absorbed into the tissues from polluted water, and the effect produced varies with the type of chemical substance, the concentration in the tissues, and the metabolism of the organism. In water that is frequently polluted, the organism is exposed to low concentrations over a varying length of time. The pollutant accumulates in the tissues and various effects can result. If the organism's metabolism destroys, or inactivates, or excretes the substance, then the tissues only receive a limited tolerable concentration that may produce no toxic effects. But if the pollutant continues to accumulate in the tissues, it may eventually reach a lethal concentration and kill the organism. Between a tolerable and a lethal concentration there is an intermediate level of toxin, which occurs as the tissue concentration is increasing, but before any toxic effects are produced. This is the threshold concentration or threshold limiting value (TLV), and it is described as the maximum concentration of a toxin that an organism may be exposed to continously, without suffering adverse effects. The TLV is used as a pollution monitoring standard, and values for chemical substances in drinking water are shown in Table 16.1.

The different toxic effects can be used to explain the results of a pollution discharge into a river. Just downstream from the outfall the water may contain no life, or very few biological species, compared to the population upstream of the discharge. The effect is caused by a high toxin concentration in the discharged effluent. This is a lethal concentration to some species, but others are able to survive because they have a comparatively high TLV, or their metabolism is tolerant to the pollution concentration. Further downstream from the outfall the effluent has become diluted, so there is a lower pollutive toxin concentration present. In this zone, ecologists state that plants and animals 'reappear', which

Table 16.1. WHO international standards for water, 1971

Substance	Maximum permissible level (g/m^3)	
	For domestic use	For drinking water
Arsenic (as As)*		0.05
Calcium (as Ca)	200.0	
Cadmium (as Cd)†		0.01
Chromium (as Cr)*	0.05	
Copper (as Cu)	1.5	
Iron (total Fe)	1.0	
Lead (as Pb)*		0.1
Magnesium (as Mg)	150.0	
Manganese (as Mn)	0.5	
Mercury (total Hg)		0.0001
Selenium (as Se)*		0.01
Zinc (as Zn)	15.0	
Ammonia		0.5
Anionic detergent	1.0	
Chloride (as Cl)	600.0	
Cyanide (as CN)*		0.05
Fluoride (at $10-12°C$)†		1.7
Mineral oil	0.3	
Nitrate (as NO_3)†		45.0
Total nitrogen (exclusive of NO_3)		1.0
Phenol*	0.002	
Polynuclear aromatic hydrocarbon (PAH)		0.2
Sulphate (as SO_4)	400.0	
Radionuclides—gross α activity		3 pCi/l
Radionuclides—gross β activity†		30 pCi/l
Coliform bacteria		Not more than $10/100$ cm^3
BOD		6
COD		10
pH	6.5–9.2	
Total hardness (as $CaCO_3$)	500	
Total dissolved solids	1500	

*Highly toxic.
†Hazardous to health.
($1 \text{ pCi} = 1 \times 10^{-12}$ curie)

means that a greater number of species are present. They are unaffected by the pollution because their tissue TLV, or their tolerable concentration, has not been exceeded. The effect of toxins upon biological life appears to depend upon the water concentration present, and the metabolism and TLV of a particular species. This is implicit in the concept of an indicator species, whose presence or absence in a stream or river is used as a rough indication of the degree of pollution present. However the continued existence of a species in a particular ecosystem is the result of many interacting factors, such as the availability of food, shelter and breeding sites; and the dissolved oxygen content, temperature and chemical composition of the water. For example, a rise of 10°C in water temperature can halve the survival time of some species, and cyanides, cresols and ammonia have a more toxic effect because the water oxygen content is lowered. Toxicity may therefore be greater in summer than at other seasons of the year.

Chemical toxins can be broadly considered under the four headings of metals and salts, pesticides, acids and alkalis, and other organic compounds such as PCBs, phenols, and cyanides. The industrial sources of the most important toxins are shown in Table 16.2. Toxic metals are often described as the heavy metals, and these include iron, lead, mercury, cadmium, zinc, copper, nickel and arsenic. Very small quantities or traces of some metals are required for normal growth and metabolism, for example copper, iron, nickel, and zinc. However, if the TLV is exceeded, then these metals may start to cause a deleterious effect, and plants and animals vary in this respect. For example, 0.3 mg/1 of zinc, 0.02 mg/1 of copper and 0.33 mg/1 of lead are lethal to sticklebacks. Plant growth is retarded by zinc concentrations of 7 mg/1 or more, but 0.5 mg/1 of copper, or 0.01 mg/1 of mercury will kill algae. Metals produce physiological poisoning by becoming attached or adsorbed on to cellular enzymes, causing inhibition of the enzymic control of respiration, photosynthesis, and growth. One of the most significant effects of metallic pollution is that aquatic organisms can absorb and accumulate concentrations in their tissues. Consequently increasing concentrations can build up in food chains and nets (see Figures 2.10 and 2.11) and they are highest in species of the secondary and tertiary trophic levels. For example, there may be up to 15 times as much mercury present in fish as in algae.

Some human populations use the higher level consumers as food, and thereby ingest the metal toxins. Several incidents of metallic toxicity are known to have been caused by people feeding on polluted marine bony and shell fish. At Toyama in Japan, wastes containing cadmium were produced from lead and zinc smelting plants, and these were discharged into the River Jintsh over a long period of time. Subsequently, the eating of fish containing a high concentration of cadmium caused a painful joint disease colloquially called 'Ouch–ouch'. This ultimately killed 100 people, and immobilized many others before the cause was discovered. Again in Japan between 1953 and 1956, waste effluent containing mercury (II) sulphate was discharged into Minamata Bay. The mercury salt was converted by marine life into organic methyl mercury, and fish provide the staple diet of the local population. Minamata disease or mercury poisoning eventually caused 43 deaths, and major disablement to another 68 people. Mercury is most toxic in the organic methyl or dimethyl form, and it is known that mercury (II) chloride used as a fungicide seed dressing can be converted by bacteria into organic mercury, both outside and inside the human body. There is evidence that mercury and other metallic concentrations are increasing in some localized areas around the coast of Britain, and in fish caught for food. Catches made in 1971 had a mean mercury concentration ranging from 0.45 g/m^3 in the river Thames and Mersey estuaries to 0.16 g/m^3 off Devon and Cornwall, compared

Table 16.2. Pollutants resulting from industrial processes

Potential Pollutants	Type of industry*												
	1	2	3	4	5	6	7	8	9	10	11	12	13
Organics													
Proteins	×	×	×	×									×
Carbohydrates	×	×	×		×	×							×
Fats and oils	×	×	×	×			×		×	×	×	×	×
Dyestuffs			×	×	×	×							
Organic acids						×							×
Phenols			×			×	×	×	×	×	×		
Detergents	×		×									×	×
Organo-pesticides		×											×
Inorganics													
Acids			×			×				×	×		
Alkalis			×			×	×	×		×	×	×	
Metals			×		×	×				×	×		×
Metallic salts				×	×					×	×		×
Phosphates, nitrates	×					×		×		×	×		×
Other salts	×		×	×	×	×							
Bleaches			×		×							×	
Sulphides			×	×				×	×				
Cyanides, cyanates								×		×	×		
Chromates				×				×		×	×		
Minerals (china clay and soil)		×				×							×

*Types are as follows:
(1) Dairying
(2) Food processing
(3) Textiles
(4) Tanning
(5) Paper making
(6) General chemical production
(7) Petro-chemical production
(8) Coking ovens
(9) Industrial oil production
(10) Engineering
(11) Metallurgy
(12) Laundry processing
(13) Agriculture

to $0.08\,\text{gm}^3$ off Iceland, and $0.11\,\text{gm}^3$ in the Irish Sea. It is alleged that there is no UK mercury hazard in fish at present, but the effect depends upon the amount of fish eaten and the cumulative quantity present in any one individual. The World Health Organisation (WHO) recommend a TLV of $0.3\,\text{g/m}^3$ of mercury for a person weighing 70 kg. It is important to know the total quantity of specific metals present in the environment, and how they become distributed in the biosphere. The behaviour of mercury is shown in Figure 16.1. Mercury contamination can enter water courses and progress via drinking water and food chains into human tissues, where it tends to concentrate

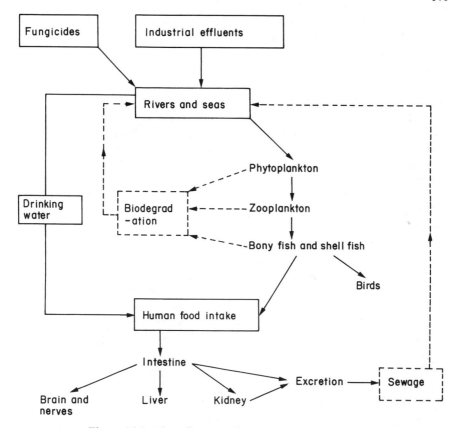

Figure 16.1. Flow diagram of mercury in the biosphere.

in certain parts of the body. Mercury can be fed back into the biosphere through the biodegradation of food chain organisms, and the discharge of sewage effluent. Metallic ions, or positively charged atoms, often accumulate in human organs. Lead and cadmium can displace calcium from bones and cause them to become brittle. Lead, cadmium, mercury, and chromium can concentrate in the liver and kidneys, and cause damage and malfunctioning of these organs. The nervous system is susceptible to concentrations of mercury, lead and copper, and the effects vary from brain damage (encephalopathy) to damage to the peripheral nerves causing uncoordinated muscular control and poor eyesight. Specific toxic effects in humans vary according to the ion type, multi-ion combinations, tissue concentrations, the frequency of dosage, and the age of the individual.

Pesticides are also a cause of water pollution. The different types of pesticides and their behaviour have been discussed in Chapter 11. These substances can enter water courses as leachates from agricultural and horticultural land, and they are present in the effluents produced from food processing plants. The most hazardous pesticides are the organochlorine compounds because of their stability and persistence in the environment. For example, DDT is present in rivers where it has an average half-life of 2.5 to 5 years, but it can persist for up to 25 years. Persistent pesticides can accumulate in food chains, and it is known that shrimps and fish can concentrate some pesticides by a factor of between 1000 to 10 000. Animals belonging to the higher trophic levels in food chains, such as birds, seals, and porpoises, may have up to 55 g/m^3 of DDT in their fatty tissues. Estimations of DDT in human body fat were made in 1964, and these varied from 3.3 g/m^3 in the UK, to 25 g/m^3 in India and Israel. Evidence of specific harmful effects in humans are difficult to assess in the long term, but there is positive evidence of DDT effects in experimental animals.

Acids and alkalis may be regarded as hazardous, because they lower or raise the pH value of water from its neutral value of pH 7. Most animals and plants will not survive in water with a pH of below 5 (acid), or above 9 (alkaline). Abnormal changes of pH in biological cells affect the activity rate of the enzymes controlling vital physiological processes. Changes in pH can also affect the action of other toxins, for example cyanides and sulphides are more toxic in acid conditions than they are in neutral or alkali solution.

Polychlorinated biphenyls, or PCBs, are by-products of the plastic, lubricant, rubber, and paper producing industries. They are stable, insoluble in water, soluble in oils, and only break down very slowly in human tissues. PCBs in discharged effluents have appeared in the tissues of fish, predatory birds, pelagic feeding marine birds, and shore mussels. These substances are lethal to these animals in quite small quantities, and in 1969 PCBs were responsible for the death of thousands of birds in the Irish Sea. Compared to DDT they have similar physiological effects, but they are more persistent in the environment, and probably less toxic in the long term. The ecological effects of other known toxins has not always been fully investigated. Cyanides are very toxic to all biological life, and probably prevent enzyme action and immobilize the nervous system in animals and humans. Chlorophenols are in use as anti-rotting agents and preservatives, and are toxic to bacteria and fish at concentrations above 1 mg/1, and to man at over 40 mg/1. A number of water pollution toxins have been designated by the WHO as potential carcinogens, capable of causing cancer in the long term. These include chromates and chromic acid, beryllium, selenium, cadmium, chlorinated hydrocarbons, some organo-phosphorus pesticides, PVC (polyvinyl chloride), nitrates, and N-nitroso compounds formed from nitrites by intestinal bacteria.

16.4 Chemical Nutrient Effects

Chemical nutrients are substances that are required by plants and animals for maintaining their growth and metabolism. In the context of water pollution, the two most important nutrients are nitrogen and phosphorus, usually present as nitrates and phosphates. Small amounts of nitrates and phosphates occur in all natural waters, and these are sufficient to maintain balanced biological growth. In lakes, over a long period of time, the nutrient levels slowly rise as a result of the biodegradation of dead organic material. This rise in nutrients is called natural ageing or eutrophication.

The concentration of nitrates and phosphates in British water courses and seas has been increasing markedly over the last 20 years. For example in 1977, the Thames Estuary contained five times more nitrate than it did in 1948, and other rivers have shown similar increases. Lakes and aquifers have also been found to have increasing concentrations of these two nutrients. Investigation has shown that there is a direct correlation between the increase in nitrates in some natural waters, and the area of arable land adjacent to them. Modern farming practices over the last twenty years have involved an increasing use of chemical fertilizers to increase crop yields. Farmers used about 1M tonnes of nitrogenous fertilizers in 1975–76, compared to 100 000 tonnes some 40 years earlier, and the present use is increasing at about 7% per annum. In the same period, 175 000 tonnes of phosphatic fertilizer were also used on British farms. All the applied fertilizers are not absorbed from the soil by growing crops, and it is probable that up to 40% of the applied nitrates enter water courses as run-off and leachate from agricultural land. Soil phosphate tends to be adsorbed, or bound to soil particles, so that probably only 20–25% of phosphate is leached into water courses. Sewage effluent contains nutrient salts because all these are not removed during the primary and secondary treatments. The quantity of phosphates present in sewage has been increasing since about 1952, when the newly developed soapless detergents began to be widely used (Figure 16.2). In the 1960s, there was a change of use from hard to soft or biodegradable detergents, but as both types contain phosphates this has not affected the steady increase, as shown. The position in 1977 was that some rivers in East Anglia had nitrate concentrations approaching the WHO recommended value of 45 mg/1 nitrogen. In the previous year, borings taken to sample the water table in East Anglia showed nitrogen levels of over 50 mg/1. In this predominantly agricultural area

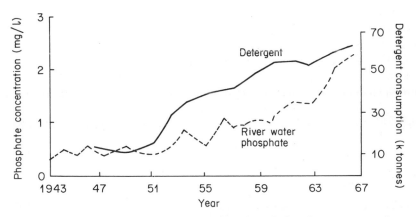

Figure 16.2. Relationship between detergent use and phosphate concentration in the River Thames water.

some water supply sources have nitrogen concentrations approaching hazardous levels, and this situation will become more serious if the agricultural industry continues its present fertilizer policy.

Phosphorus is required for the photosynthetic process in plants, for respiration, and the production of nuclear DNA in all plant and animal cells. Nitrogen is an essential constituent of proteins, which are present in cell cytoplasm, the nucleus, and enzymes in all biological organisms. Increased concentrations of nitrates and phosphates in water produce the overall effect of an increase in the rate of growth. For example, unicellular green and blue-green algae, and floating filamentous algae such as *Cladophora* or 'blanket weed' grow rapidly in eutrophic conditions. A dense, slimy surface layer produces the so-called algal 'bloom' condition, particularly in slow-moving waters in the spring and summer. This bloom reduces light penetration and restricts atmospheric reoxygenation of the water. The effect is to cause adverse conditions for river and canal craft navigation, and for swimming, bathing, and fishing. The dense algal growth eventually dies, and the subsequent biodegradation produces an oxygen deficit which can result in obnoxious anaerobic conditions. Well known examples of the severe effects caused by eutrophication have been described for Lough Neagh in Northern Ireland, and lakes Erie and Ontario in Canada, but minor effects have occurred in some waters in GB. The growth rate of animals is speeded up, especially in fish. In fact, the concept of fish farming at present being experimentally investigated, involves the artificial enrichment of enclosed areas of water by adding nutrients. In relation to humans, there is concern about excessive nitrate concentrations in drinking water abstracted from rivers, aquifers, and bore holes. In 1978, it was stated that more than 100 public water sources in GB had intermittent or continuous levels of nitrate near to the WHO limit. This hazard can be reduced by dilution of the source water, or by a temporary cessation of abstraction. Nitrates are taken into the body in food and drink and can cause disease. Some bacteria in the intestine can reduce nitrate to nitrite, and when this enters the blood the nitrite becomes attached to the red blood pigment or haemoglobin, to form a complex called methaemoglobin. This produces a reduction in the oxygen carrying capacity of the blood, and a condition called anoxia, or methaemoglobin anaemia, or the blue baby disease. It is particularly severe in young babies and may be fatal. It has been suggested that nitrites can be further converted to amines and nitrosamines in the human body, and these organic chemicals are a possible cause of gastric cancer. The concentration of nitrates in natural waters is increasing, and this could become a potential health hazard in the future.

16.5 Micro-organism Effects

Wastes that are discharged into water contain pathogenic organisms that are capable of transmitting human diseases. Some bacteria are water borne, and these include types responsible for causing cholera, typhoid fever, bacillary dysentery, and gastroenteritis. Viruses are also found in water, including strains which cause poliomyelitis, infective hepatitis, and echo and coxsackie fevers. The infective egg and larval stages of some animal parasites also occur in water, including the round worm (*Ascaris*), and the beef and pork tape worms. All these types of organisms occur in faeces, and so are present in sewage and farm slurry. The primary and secondary sewage treatment probably removes from 50% to 95% of all bacteria and some viruses, but this is incidental to the treatment of the organic matter. Usually some pathogenic organisms are present in sewage effluent, and when this is discharged they enter water courses. Farm yard slurry can contain parasitic stages and organisms responsible for salmonella food poisoning and brucellosis. When slurry is spread on land without proper precautions having been taken, run-off and ground water drainage into rivers and streams can cause micro-organism pollution.

Water Authorities are responsible for safeguarding public water supplies. At present, sewage effluent is tested only for the presence of *Escherichia coli* (*E. coli*), which is a non-pathogenic intestinal bacterium. The so-called coliform test of water purity is carried out to indicate the presence or absence of faecal matter present in water. This test does not show the presence of pathogenic bacteria, and few tests are at present in use for viruses, although suitable monitoring techniques are under investigation. Therefore those inland and coastal waters that receive sewage discharges are a potential health hazard. However there is little evidence to show that people are infected or at risk when they swim or bathe in these waters. Drinking water supplies that are abstracted from rivers are safeguarded by the pre-water supply chlorination treatment. However it is not possible to carry out chlorination of sewage effluent before it is discharged into water. This is because some free chlorine would enter rivers, and very low concentrations, such as 0.03 mg/l, are lethal to fish and interfere with biodegradation, which is essential for the self purification of rivers.

16.6 Radionuclide Effects

The increasing development of nuclear energy in Britain is producing more radioactive waste to be disposed of into the environment, and it contains various radionuclides with long half lives. At present low and medium activity wastes are either stored on land or disposed of at sea. For example in 1976, there were 12 000 m^3 of solid wastes stored on land in the UK, and this contained nearly half a tonne of plutonium. Other wastes are sealed into containers and dumped into the North Atlantic at a depth of 4500 metres, at a location 900 km SSW of Lands End. In 1978, the DOE stated that about 66 000 tonnes of packages of solid low activity waste had been dumped at sea since 1949, and the present scale of dumping was about 7000 tonnes per year. This solid waste is dumped subject to the provisions of International Conventions, the IAEA recommendations, and the OECD Nuclear Energy Agency supervision (see Chapter 25). When these radioactive wastes are dumped at sea they are safe, but there is concern regarding what could happen in the future. The corrosive action of sea water and the effects of natural forces may eventually cause damage to, or leakage from the containers.

If this should happen, the escape and spread of radioactive nuclides would be uncontrollable, and some material would enter the marine ecosystem. Low level activity liquid wastes are discharged by pipeline into coastal waters. In the UK, 75% of the total waste is produced at Windscale in Cumbria, and liquid effluent is discharged into the Irish Sea. This contains such radionuclides as ruthenium-106, strontium-90, cerium-144, caesium-137, and various plutonium isotopes. These substances settle on to bottom sediments and become adsorbed on to the mineral particles. It has been suggested recently that the radionuclides may not remain static at the discharge or dumping sites. Clearly isotopes that leaked into the sea could be circulated by storms and ocean currents and eventually reach coastal waters. Here the radionuclides could follow two environmental pathways. Above the shore high tide level adsorbed particles could become air borne as dusts and aerosols. These could be breathed in, or coat vegetation, and so enter the bodies of animals and humans. Alternatively, chemical changes in the sediments could release or desorb the radioactive particles, with the result that they would become suspended, washed ashore, and again be absorbed by plants, animals, and people. In both cases the environmental pollution path involves food chains and nets. It is known that caesium-137 can become concentrated in bottom living fish such as plaice, dabs, and skates. Man uses members of the higher trophic levels as food, for example fish, shellfish, and crustaceans such as crabs, lobsters, and shrimps. A MAFF report in 1972–73 stated that some of the nuclides discharged from Windscale produce local concentrations in the Ravenglass estuary. However salmon fishermen, who use the estuary, had absorbed only 9% of the recommended ICRP maximum permissible dosage. Analysis of fish caught in the Irish Sea showed that they contained less than 0.1% of the ICRP dosage. Therefore in the UK there appears to be no evidence to suggest that the discharging of low level activity wastes into coastal waters is causing a health hazard at present. However, research work and close monitoring is required in the future, to discover the environmental pathways of radioactive wastes and the chemical changes that occur in polluted sediments.

CHAPTER 17

Future Developments

Water pollution like other forms of pollution is caused basically by man's inability to dispose of waste in ways that do not change the natural balance of the environment. The most serious pollution effects are caused by waste chemical substances entering and accumulating in ecosystems and food chains. Biological life and human health can be affected, and the quality of the environment becomes degraded.

Freshwater is a valuable natural resource essential for man's personal use and food production, and indispensible for many industrial processes. The UK consumption of water is increasing at just over 2% per year, and the cheapest source of supply is natural water courses. But the rivers from which public and industrial supplies are abstracted are also used for the disposal of waste effluents. In the future, the continued pollution of rivers by insufficiently treated effluents can only result in less water being available, and increasingly expensive pre-water supply treatment. This is one of the present dilemmas in respect of water supplies. Natural waters, whether rivers, lakes, canals, or the sea are also a source of aesthetic enjoyment, recreation, and transport for millions of people.

Increasing pollution affects the quality of enjoyment of the environment and may endanger the health of water recreationists. The greater mobility of the population, and the increasing amount of non-working time they possess is creating more demand for the leisure use and enjoyment of water. This trend will probably increase in the future. For all these reasons it is essential to reduce the extent of water pollution both now and in the future.

The production of waste cannot be eliminated, and the disposal of waste is a necessity. It is generally accepted that the diversity and quantity of wastes produced will continue to increase in the future. If these two factors could be reduced, then disposal problems and the pollution caused might be reduced. In other words, the problems of waste should be tackled at their source; before, and not after wastes have been discharged into the environment.

Large volumes of sewage effluent and sludge are produced containing organic matter and many inorganic salts and metals. Smaller quantities of industrial effluent are produced with a wide range of potential pollutants. Both types of waste require treatment using expensive plant, that must be continually up-dated to keep pace with advancing manufacturing and production technology. A concerted effort should be made to determine how much of this waste can be extracted and re-used, and what amount remains which must be disposed of into the environment. This will require research and a detailed study of the economics involved. For example, the organic content of sewage and farm animal wastes is highly polluting, but more use could be made of it as a processed fertilizer and soil conditioner. There is evidence that many wastes containing organic matter, which result from human excreta, silage, textile and food processing, paper making, and spent lubricants, could be re-utilized. Such wastes have been used in experimental microbiological processes to produce alcohol and proteins. Sewage contains inorganic compounds, many of which are not extracted by tertiary treatment. In 1971 it was estimated that the population of GB produced sewage containing 500 tonnes of saline nitrogen, and 30 tonnes of phosphates per day. These are useful compounds that are causing increasing pollution in rivers, and could be extracted and utilized. Industrial wastes contain numerous solvents and dissolved compounds that could be recovered and recycled back into production. Studies need to be carried out to establish the economic benefits of recycling compared to the costs of treatment and disposal handling. At present some commendable efforts are being made through the Reclamation Industries Council, and bodies such as the National Industries Materials Recovery Association (NIMRA), the Chemical Recovery Association (CRA) and the DT Warren Springs Laboratory. More reclamation of wastes will reduce the use of natural resources, and also reduce the quantity and complexity of waste disposal. Some chemical wastes are highly polluting and cannot be easily treated to make them innocuous. For these, and other types of waste, more use should be made of disposal by incineration. The process is expensive and requires specialist operation, but regional plants could be established, and use made locally of the heat energy produced.

Irrespective of any future improvements in waste management there will still be a need to improve the control of wastes discharged into water courses. Monitoring is essential to determine the identity and levels of pollutants and the effectiveness of discharge controls. The present national river pollution surveys have provided a broad review of inland water quality, and there are revised and more detailed proposals for future surveys from 1979. However there is a need to establish a national monitoring network, to measure the levels of a wide range of specifically hazardous substances such as heavy metals, nitrates, phosphates, PCBs, pesticides, and carcinogenic hydrocarbons, to replace the scattered sampling carried out at present. In the future, more attention should be directed towards coastal waters pollution. From 1980, the Water Authorities will be required to monitor and control all discharges into coastal waters under part 2 of the Control of Pollution Act 1974. This will be an important first step in the improvement of the coastal area pollution that has hitherto been much neglected. There is concern that the continued discharge of low level radioactive effluents into coastal waters will result in increasing concentrations of long-life isotopes in the ecosystem. There is also evidence that radioactive wastes discharged from Windscale in Cumbria can travel over 800 km to the West German coast. In future, monitoring of isotope concentrations in sea foods and marine food chains in the vicinity of discharges should be carried out intensively. In addition, on a wider scale, both on-shore and off-shore monitoring stations should be established to measure the wider oceanic spread of radioactivity.

Improved monitoring organization and procedures will only help to reduce water pollution, if the information obtained is used to implement better control and treatment of wastes. There is a future need to improve all effluent plants operated by the WAs and industry. Sewage treatment should be made more efficient by including more tertiary treatment processes to remove inorganic pollutants. The WAs also need to increase their overall treatment capacity, and to reduce significantly the quantity of untreated and partially treated effluent released into inland and coastal waters. The discharge consent conditions for industrial effluents should be made more stringent in the future. This could result in increased and improved on-site waste treatment, or the storage and non-release of some very hazardous pollutants into rivers. In the medium and long term future, the design of production processes and plant should always include consideration of the resulting wastes and their method of disposal.

All the above proposals could produce a decrease in water pollution, but they would be costly to implement. The economics of pollution control and abatement need to be more closely studied. It is relatively easy to estimate the cost of waste treatment amd disposal, but at present there is no acceptable method of assessing the financial cost to both industry and society, based upon achieving desirable environmental quality objectives. The present policies in operation to reduce pollution are all constrained by the need to carry them out at minimum cost to industry and the national economy.

References

Besselievre, E. B., *The treatment of Industrial Wastes*, McGraw Hill, 1967.

Best, G. A., and Ross, S. L., *River Pollution Studies*, Liverpool University Press, 1977.

Calleley, A. G., Forster, C. F., and Stafford, D. A. (Eds), *Treatment of Industrial Effluents*, Hodder and Stoughton, 1977.

Department of the Environment, *Background to Water Reorganisation in England and Wales*, HMSO, 1973.

Department of the Environment, *River Pollution Survey of England and Wales*, HMSO, 1972 and 1976.

Department of the Environment, *Taken for Granted—Report on Sewage Disposal*, HMSO, 1970.

Gilpin, A., *Dictionary of Environmental Terms*, Routledge and Kegan Paul, 1976.

Gowan, D., *Slurry and Farm Waste Disposal*, Farming Press, 1972.

Haynes, H. B. N., *The Biology of Polluted Waters*, Liverpool University Press, 1971.

Holister, G., and Porteous, A., *The Environment: A Dictionary of the World Around Us*, Arrow, 1976.

House, J. W. (Ed.), *The UK Space—Resources, Environment and the Future*, Weidenfeld and Nicolson, 1973.

Imhoff, K., Muller, W. J., and Thistlethwaythe, D. K B. (Eds), *Disposal of Sewage and Other Water-borne Waste*, Butterworth, 1971.

Loftus, T., *The Last Resource*, Penguin, 1972.

Massachusetts Institute of Technology, *Man's Impact on the Global Environment*, MIT Press, 1971.

Mellanby, K., *The Biology of Pollution: Institute of Biology Study No. 38*, Arnold, 1972.

Neal, A. W., *Industrial Waste*, Business Books, 1973.

Open University, *Water: Origins and Demand, Conservation and Abstraction, PT272 Units 3 and 4*, OU Press, 1975.

Open University, *Clean and Dirty Water, Water Analysis Standards and Treatment, PT272 Units 5 and 6*, OU Press, 1975.

Overman, M., *Water*, Aldus Books, 1968.

Rothman, H., *Murderous Providence*, Rupert Hart-Davis, 1972.

Royal Commission on Environmental Pollution, *First Report*, HMSO, 1971.

Royal Commission on Environmental Pollution, *Third Report*, HMSO, 1972.

Royal Commission on Environmental Pollution, *Fourth Report*, HMSO, 1974.

Royal Commission on Environmental Pollution, *Sixth Report*, HMSO, 1976.

Simmons, I. G., *The Ecology of Natural Resources*, Arnold, 1974.

Tebbitt, T. H. Y., *Water Science and Technology*, John Murray, 1973.

Walker, C., *Environmental Pollution by Chemicals*, 2nd Edn, Hutchinson, 1975.

SECTION V

Noise

CHAPTER 18

Sound, Hearing, and Noise

For a long time many individuals have accepted noise as a part of their environment. Attitudes have varied from indifference—'it can't be helped', to bravado—'it doesn't worry me', or fatalism—'if you work or live here you must expect it'. In these situations individuals have not considered noise as a cause for concern. However, over the years the general incidence of noise has been increasing. The development of the steam engine, the petrol engine, and technological machinery in industry, all contributed to an increasingly noisy environment in the nineteenth century. This has been further exacerbated in the twentieth century by the diesel engine, the turbo-prop and jet engines, the increasing use of faster industrial production machinery, construction site machinery, and the increased volume of road traffic. The Committee on the Problem of Noise, Final Report 1963 (Wilson Report), stated that of 1400 people interviewed about noise in 1948, 23% said they were disturbed by noise, but by 1961 the proportion had doubled to 50%. However, gradually over the last decade a growing concern about noise has developed. Noise is now regarded by many people as a pollution component that contributes to a deterioration of the environment. Noise is now accepted as a potential hazard to health and communication. Government recognized this by the inclusion of noise in various statutes brought into operation in 1969 and the following years.

18.1 Sound and its Characteristics

Sound consists of wave motion in an elastic medium. The medium may be air, water, or solids such as metals, plastics, wood, bricks, and concrete. Sound waves travel through the medium from the source where the sound is produced, to the recipient or listener. Sound waves consist of variations in pressure, or oscillations of the medium in which they travel. The rate of the oscillations is called the frequency of the sound, and is measured in cycles per second or hertz (Hz). The frequency determines the pitch of the sound received by the listener. High pitched sounds have high frequencies and these are more disturbing to the individual than low frequencies, because of the varying sensitivity of the human ear. Sound pressures consist of small variations in the sound media, and as they increase, the sound becomes louder. Sound pressure is measured in newtons per square metre

(N/m^2). The third component of sound that can be measured is sound intensity. This is the amount of sound energy that flows through a unit area of the medium in a unit time interval. Sound intensity is measured in watts per square metre (W/m^2). Sound intensity decreases as the square of the distance between the sound source and the listener. An illustration of this is the sound pressure measurement of traffic noise on a motorway. At a distance of 7 m away, the sound pressure is 78 decibels (dB), and at 14 m it is 75 dB, but at 28 m it is 72 dB.

Sounds are often described in terms of loudness, but this cannot be measured as is the case with frequency, sound pressure, and sound intensity. Loudness can be described as a listener's auditory impression of the strength of a sound, and it is therefore a subjective judgement. Loudness is not synonymous with sound pressure, because a sound of 60 dB constant pressure is louder to the listener if the frequency is 1000 Hz, and quieter when the frequency is 100 Hz. Alternatively a sound of 20 Hz must have a sound pressure of 80 dB to be audible, but a sound of 2000 Hz is audible at a pressure of 20 dB.

18.2 The Measurement of Noise

This is not a simple operation for a number of reasons. As previously stated, the judgement of what sound quality and intensity is called noise is subjective. Some people can tolerate higher noise levels than others, and they can become accustomed to continuous high levels of noise. The human ear is not equally sensitive to different frequences, sound intensities, and pressures. Also noise is often not produced at a constant frequency, nor is it continuous over a specific period of time. It is often intermittent high peak noise levels and their periodicity, for example road traffic and aircraft noise, that cause the greatest annoyance.

The two most important measurements of noise are sound pressure and sound intensity. These are measured in different units, and the scale of magnitude is different and very large in relation to noise. The scientific acoustic unit in common use is the decibel (dB). It is not an absolute physical measurement unit comparable to the gram, volt or metre, but it is a ratio expressed as a logarithmic scale relative to a reference sound pressure level. The decibel is defined as:

$$\text{sound intensity level} = 10 \log_{10} \cdot \frac{\text{intensity measured}}{\text{reference intensity}}.$$

The logarithmic scale enables the wide range of sound pressure and intensities to be conveniently accommodated. The relationship between the three measurements is shown in table 18.1. The reference intensity used is the threshold of hearing, which is a sound that can be first heard at a sound pressure of 2×10^{-5} N/m^2 or a sound intensity of 10^{-12} W/m^2. It is very important to realize that a doubling of sound pressure produces an increase of 6 dB, and a doubling of sound intensity produces an increase of 3 dB, which is implicit in the logarithmic scale for sound measurement. The dB scale is not satisfactory for measuring noise, because this needs to be related to the human ear frequency response, and the environmental circumstances in which the noise is produced. The human ear

Table 18.1. Intensities, sound pressures, and decibels (dB) for sound in air, at room temperature, and sea level pressure

Intensity (W/m^2)	Pressure (N/m^2)	dB	Sound source
100 M	200 000	200	Saturn rocket take-off
1.0	20	120	Boiler makers' shop
10^{-2}	2.0	100	Siren at 5m
10^{-4}	2×10^{-1}	80	Heavy machinery workshop
10^{-6}	2×10^{-2}	60	Normal conversation at 1m
10^{-8}	2×10^{-3}	40	Public library
10^{-10}	2×10^{-4}	20	Church
10^{-12}	2×10^{-5}	0	Threshold of hearing

has varying sensitivity to different frequencies as already quoted. Broadly, it is less sensitive to lower than to higher frequencies, and has a limited range. Consequently noise measuring meters have been designed with networks that compensate for this, by reducing the response to low and very high frequencies. These meters record the dBA scale which is in common use for the measurement of general noise levels. This scale has the convenient feature that a doubling of sound frequency is equal to an increase of 10 dBA.

The dBA scale is in general use, but it is not sufficiently refined to take account of peak noise levels, the duration of noise exposure, and the quality of noise that are features of specific environmental noise situations. Consequently other noise indices have been devised based upon the dBA scale, but having built-in refinements, or weightings, to provide a calculated single figure index.

L_{10} (18 hour) Index

This index is used for road traffic noise measurement, and it has been adopted for use in UK noise legislation. The index is expressed in dBA, and it is the arithmetic average hourly values of the level of noise exceeded for 10% of the time, between the hours of 0600 and 2400 on any normal weekday. It takes account of the peak noise values, and variability of noise depending upon the type of vehicle and the traffic density.

Perceived Noise Level (PNdB)

This index is used for measuring aircraft noise levels in the UK. It is equivalent to the dBA scale plus 13. An increase of 10 PNdB is equivalent to a doubling of the noise.

Noise and Number Index (NNI)

An index used to express the extent of disturbance from aircraft that produce intermittent peak levels when they take-off and land. The NNI indicates the average maximum perceived noise during the passage of successive aircraft, and the number of aircraft heard during a specified daylight period. The index starts at 67 dBA (80 PNdB), and an increase of 5 NNI arises from an increase in the average peak noise level of 5 PNdB for a given number of aircraft, or from a doubling of the number of aircraft with a given peak noise level.

Effective Perceived Noise Level (L_{EPN})

This index is recommended for aircraft by the International Civil Aviation Organisation (ICAO) as the standard for use in noise evaluation in aircraft noise certification. The index is based upon the PNdB scale, takes account of the peak frequency jet aircraft noise and the duration of aircraft flyovers, and is accepted by the UK.

Corrected Noise Level (CNL)

This is used in the UK for measuring noise disturbance in industrial premises. It is expressed in dBA and is used as a British Standard. The index takes account of the intensity, loudness, intermittency, duration, and tonal and impulsive character of the noise. The CNL was derived for the prediction of complaints, and it is complicated by dependence upon subjective qualitites. The Noise Advisory Council Report 1971 propose a maximum CNL of 75 dBA by day, and 65 dBA by night.

Equivalent Noise Level (L_{eq})

This is recommended by the International Organization for Standardisation (ISO) for measuring and rating noise in residential, industrial, and traffic areas. The index is an equivalent or mean energy level over a specified period of time. An L_{eq} of 90 dBA indicates a steady noise level over a whole period of time, or a noise level steady at 93 dBA for 50% of the time, and no noise for the remaining period, and so on. The index is in use in a number of countries for specified day, evening, and night periods.

In addition to the above noise indices that are in common use, there are also the Noise Exposure Forecast (NEF) and the ICAO recommended Weighted Equivalent Continuous Pereceived Noise Level (WECPNI), both devised for aircraft noise, and also the Traffic Noise Index (TNI) for road traffic. The fact that there are ten different methods of expressing noise, in addition to the decibel scale, is very confusing to the non-expert, and makes comparison of the scales impossible in most cases. The Noise Advisory Council reviewed these noise indices in 1975, and recommended that one unified scale, the dBA scale 'should be used for measurement of noise levels from all sources for the purposes of environmental noise assessment'. No unified scale was recommended for measuring the noise environment, but the L_{eq} level was proposed with indices based upon it.

18.3 The Ear and Hearing

The human ear receives sound waves, and these set up oscillations in the tympanic membrane or ear drum. These oscillations cause sympathetic movements of the three ossicles or small bones in the middle ear behind the ear drum. The oscillations then pass through fluid in the inner ear to the auditory nerve, and on to the brain. In the brain, the oscillations or sounds are identified and interpreted. The brain is able to select mixed sounds into different categories, for example it can distinguish speech from background noises, and if desired it can consciously suppress unwanted sounds, depending upon their intensity. The ear is able to analyse sounds into frequency components, and the range of an 18 year old person with perfect hearing is between 20 Hz and 20 000 Hz. Normally, hearing is most acute in the frequency range 2000 to 5500 Hz, but it falls off rapidly below 200 Hz and above 10 000 Hz. However the ear's sensitivity range varies considerably from person to person. As people age, they experience a progressive hearing loss of high frequency sounds which is called presbycousis. The usual effect is a limitation of the maximum high frequency range to 16 000 Hz. This occurs noticeably after 40 years of age, and is more evident in men than women, perhaps because of the effects of occupational noise. It is normal for a 70 year old person to have a 60 dB loss of sensitivity to a 12 000 Hz tone, and a 10 dB loss to a 500 Hz tone, compared to a 30 year old person. The ear is potentially liable to damage if it receives high intensity noise. To give some measure of protection the ear has two reflex actions. If the ear receives a noise of over 90 dB for longer than about 10 milliseconds, the tympanic membrane contracts. This effectively reduces the sound energy received, reduces the movement of the ossicles, and hence the sound transmission to the inner ear. This behaviour is called the aural reflex. A different mechanism operates for sound over 140 dB. In this case, the middle ear muscles change the direction of movement of the ear ossicles, and this causes a sudden decrease in the noise intensity reaching the inner ear. Both these protective reflexes have only a limited effect against loud noise, and they are only effective for very short durations.

CHAPTER 19

Effects of Noise Pollution

19.1 Hazardous Effects

There is no doubt that noise can damage the ear and cause temporary or permanent noise-induced hearing loss, depending upon the intensity and duration of the sound level. The most commonly occurring ear damage is caused by continuous periods of high intensity noise. If the ear receives a noise level of over 90 dB in the mid-high frequency range for more than a few minutes, then the auditory sensitivity is reduced. The effect is called the temporary threshold shift, and can result in a sound of 4000 Hz frequency requiring a noise level of 20 dB to be heard, instead of 5 dB under normal conditions. This effect is shown in Figure 19.1. If the noise of 90 dB ceases, then the temporary threshold shift disappears in, say, 30–60 minutes, and the normal hearing sensitivity is restored. The length of the recovery time depends upon the noise level and the duration. For example, an exposure to a 100 dB noise level in the 1200 to 2400 Hz frequency range for 100 minutes could produce a 30 dB shift, and require 36 hours to recover. Recovery is dependent upon the fact that the excessive noise level is of comparatively short duration, and so no permanent damage usually results. In some factories and other work situations, workers are constantly subjected to high noise levels. Under these conditions the auditory threshold shift is no longer temporary, and progressive hearing deterioration results. This effect is also shown in Figure 19.1, where it is seen that the degree of hearing loss is reduced sensitivity in the 500 to 8000 Hz frequency range. After years of such occupational noise exposure the worker suffers chronic hearing loss.

Besides progressive hearing loss, there may be instantaneous damage or acoustic trauma. This is usually caused by very high intensity impulse type noise, that can result from an explosion, or sudden excessive noise of about 150 dB or more. Sonic booms or over-pressure from supersonic air liners are impulse noises, which some people think can damage the ears. There is no evidence for this at present because sonic booms seldom exceed 100 N/m^2, and it requires a peak energy of 35 000 N/m^2 to rupture the tympanic membrane. Low intensity impulse noises can also cause ear damage.

Excessive noise does not invariably cause ear damage, but alternatively it may cause pathological or psychological disorders. Pathological effects can result from particular noise frequencies causing vibration or resonance in materials or

192

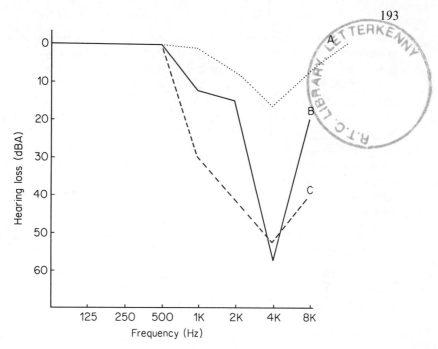

Figure 19.1. Typical loss of hearing resulting from weaving. Curves: A, after a short period of exposure to noise; B, after 20 years' exposure; C, after 35 years' exposure. (After *Noise*, 2nd edn, by Rupert Taylor (Penguin, 1975).)

people's bodies. High frequencies, or ultrasonic sound above the normal audible range, can affect the semi-circular canals of the inner ear and cause nausea and dizziness. Alternatively, very low frequency noise can cause resonance in the body organs, producing the effects of decreased heart beat, variations in blood pressure, and breathing difficulties. Mid-audible band frequencies are known to cause resonance in the skull, and so affect the brain and nervous system with consequent effects upon thinking and co-ordination of the limbs. Noise and physical vibrations from hand-held tools can have varying effects, often called white fingers, or dead hands, or Reynaud's phenomenon, or pneumatic drill disease. Pain, numbness, and cyanosis (blue colouration) of the fingers can result from moderate vibration, but damage to bones and joints in the hands with swelling and stiffness can be caused by severe vibration. The pathological effect of noise upon particular individuals is very variable, and not yet fully understood. In some cases it seems likely that physical vibrations affect the nervous system and the ear, causing hearing disturbance.

Non-pathological or psychological noise effects are also variable and very difficult to measure. The mildest effect is often physical and mental fatigue and lack of concentration. This effect is important in industrial situations, because it results in lowered efficiency, a reduced work rate, increased absenteeism, and a higher potential for accidents and injuries. In the non-work environment, psychological noise effects impinge upon sleep. Every person requires a specific minimum period of regular sleep of good quality, to maintain a healthy body and mind. The intensity or depth of sleep and its duration is important, because it is during this phase of the sleep period that the most recuperative benefit is obtained. Low frequency noise of say 50 to 60 dBA can affect the higher centres of the brain, and cause an alteration to the normal sleep pattern and prevent deep sleep. In young and middle aged people this noise may not wake them, but in older people it can wake them and prevent a return to sleep. Again the effect of noise upon sleep varies greatly, and a noise level of 40 dBA may wake some people, but for others the required level may be as high as 70 dBA. Noise can also affect verbal communication upon which we all depend, whether in the work, domestic, or social environments.

19.2 Hearing Loss and Noise Limits

It is unfortunately true today that many people work and live in environmental conditions where the noise level is not obviously hazardous. Nevertheless, over long periods of time, they do suffer progressive hearing loss over and above that

Table 19.1.   Degree of hearing loss and its effect

Effect	Average hearing loss (dB)
No real difficulty in hearing	less than 25
Difficulty with hearing soft speech	25 to 40
Difficulty with hearing normal speech	40 to 55
Difficulty with hearing loud speech	55 to 70
Only shouted speech understood	70 to 90
Unable to hear even amplified speech	90 or more
Hearing aid required	50 or more

Table 19.2.   Maximum permissible noise levels

Situation	dBA
Road traffic near residential areas	70
Noise on building or construction sites	70
Ear protection should be worn	85
Factory work for an 8 hour day, 5 days per week	90
Prolonged noise causing permanent damage	100
Threshold of pain—duration of 30 secs maximum	120
Absolute maximum with ears unprotected	135
Maximum for impulse noise	135
Maximum for instantaneous noise	150
Absolute limit with ears protected	150
Ear drum rupture	180
Lung damage	194

causd by presbycousis. This gradual effect, and the degree of hearing loss, has been summarized by W Burns in '*Noise and Man*' (see Table 19.1).

Alternatively there are many environmental situations where the noise level is recognized as sufficiently high as to be hazardous to hearing. For some of these general situations, maximum permissible noise levels have been recommended or recognized. (Table 19.2).

None of these limits have any statutary force, but there are prescribed limits for specific conditions that will be described later. To assess the hazards of specific noisy conditions Table 19.2 should be compared to Tables 21.1 and 22.1, which show the noise levels in a wide range of environmental situations.

CHAPTER 20

Transport Noise

Noise is slowly becoming recognized as an unjustifiable interference and imposition upon human comfort, health, and the quality of modern life. Noise occurrence can be roughly divided into two broad categories. Hazardous noise is usually associated with industrial work situations, and employers and employees have had a considerable degree of tolerance to this type of noise in the past. The attitude has been a reluctant acceptance of noise in a captive work situation, where particular manufacturing processes have meant an imposed noisy environment. Alternatively, disturbance noise is often described as the environmental noise to which people are subjected outside their place of work. This type of noise is very individually subjective in its effect, but it is causing increasing concern to more and more people, who are becoming increasingly less tolerant to it. A Market Research International Survey 1972 reported that in a sample population, 12% of people considered noise as the prime environmental hazard in the UK. Within this 12%, there were 72% of people who stated that noise was getting worse, and more than 50% were seriously disturbed by noise. The Association of Public Health Inspectors stated that complaints about noise nuisance in local neighbourhoods were increasing by about 10% per year, with road traffic and aircraft noise the targets of universal criticism. The standing Royal Commission on Environmental Pollution 4th Report 1974 predicted that 29 M people in the UK, or 50% of the population, will be subjected to unacceptable traffic noise by 1980, assuming the present noise control conditions. This statement can be compared to a Department of the Environment estimate in 1976, that 20M people live adjacent to roads with unacceptable noise levels.

20.1 Road Traffic Noise

Noise from road vehicles produces disturbance to more people than any other noise source, and this has been increasing over the last decade for a number of reasons. The total number of road vehicles, and hence the density of road traffic, is steadily increasing. In 1967 there were 14M licensed road vehicles, but 9 years later in 1976, the total was 17.8M. The average increase in licensed road vehicles of all types is just under 2% per year. In 1976, about 10% or 1.8M vehicles were used for goods haulage and these lorries are increasing in size, especially for container traffic. In 1967, 133 000 licensed vehicles were over 5 tonnes unladen weight, but by 1976 the total had risen to 229 000, and the number of lorries of over 8 tonnes trebled over this 9 year period. To these totals must be added the non-UK registered continental lorries. The average traffic density of all types of vehicles on UK roads is also increasing. The estimated density in 1967 was 184.8 vehicles per kilometre, and this increased to 259.4 in 1977. These trends mean that the increased numbers and density of all road vehicles inevitably increases the road traffic noise.

One of the most important causes of noise on the roads is the traffic speed. The faster the traffic travels, the greater the noise volume, and modern road development policy is encouraging higher speeds. Since the late 1950s, successive governments have constructed six-lane motorways at an increasing rate, so that by 1977 there were 2286 km of motorway in use in the UK. This development has allowed large volumes of traffic to travel at sustained speeds up to 113 km per hour for cars, and 80 km per hour for heavier vehicles. The construction of new motorways has been paralleled by improvements in the standards of other roads. Trunk roads have become dual carriageways, urban bypasses have skirted towns, and urban ring roads and elevated freeways have increased the speed of the traffic flow in and around built-up areas. Speed limits have risen from 48 km per hour to 64 or 80 km per hour on many roads. All these road developments have invariably been in urban communities or adjacent to them, where roads pass through commercial and residential housing areas, so producing increased traffic noise to all the inhabitants.

Road traffic noise fluctuates according to a number of operating factors. Noise is produced by all vehicles from the gear box and exhaust system. Heavy vehicles also produce rattles, squeaks and vibrations according to the degree of loading

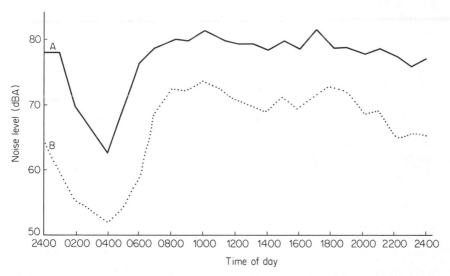

Figure 20.1. Noise levels on the M4 motorway taken at 6 m from the nearest carriageway. Curves: A, 10% of total vehicles; B, 90% of total vehicles.

and age. All vehicles produce more engine noise at faster speeds, and a doubling of engine speed can increase the noise level by 13 dBA. Also tyre noise increases with speed, and a wet road can increase the noise by 10 dBA. Generally a heavy goods vehicle produces twice as much noise as a private car, or to a motorway observer the noise from one heavy lorry equals that from 10 passing cars. The actual pattern of traffic noise on a main road is complex. There is a general noise level as long as any traffic is moving, and this varies with the traffic density and the time of day (Figure 20.1). It is common on urban roads to have regular distinctive traffic peaks in the morning and evening as people travel to and from work. Superimposed on this general level are peak traffic noise levels of a few seconds duration when individual vehicles are passing any given point. These individual noise peaks vary according to the size, type, and speed of the vehicle. For example, heavy diesel-engined goods vehicles are the noisiest vehicles on the roads today, and electrically propelled vehicles are the quietest.

Traffic noise can be accurately measured in dBA, and noise levels are often expressed on the L_{10} (18 hour) index. Some typical examples are given in Table 20.1. The actual noise limits in force vary in different European countries. A comparison between the UK and the EEC is made in Chapter 28 on noise control legislation. In the UK, the noise limit for cars coming into use after 1.11.1974 is 80 or 82 dBA, according to whether they are petrol or diesel-engined models. In France, the limit is 83 dB for cars, and in Switzerland it is 70 dB for cars during daylight, and 60 dB during the hours of darkness. In the UK, the noise limit for heavy vehicles over 200 b.h.p. is 89 dBA from 1.11.1974, and yet as far back as 1963 the Wilson Committee proposed a limit of 85 dB for heavy vehicles.

Various estimates and surveys have been of the extent to which road traffic noise disturbs people, and the noise levels that they experience. A Transport and Road Research Laboratory Report 1970 produced the information shown in Table 20.2.

Table 20.1. Typical road traffic levels

Situation	Average traffic speed (km per hour)	Intervening ground	L_{10} (18 hour) index (dBA)
Residential road, parallel to busy main road	48 (30mph)	Houses	60
18.3 m (60ft) away from busy main road in residential area	48	Paved area	70
18.3 m from motorway with high volume of heavy vehicles	96 (60mph)	Grass	80
3 m (10ft) from main road in residential area	48	Paved area	80

Table 20.2. TRRL noise levels and UK residents, 1970

L_{10} noise level	Population in millions			
	Urban	Rural	Total	%
70 dBA or more	8.5	1.2	9.7	17.8
65 to 70 dBA	12.5	1.7	14.2	25.6
Less than 65 dBA	24.3	6.2	30.5	56
Total	45.3	9.1	54.4	—
	(83%)	(17%)	—	

The Noise Advisory Council recommended an L_{10} index of 70 dBA as the absolute limit of road traffic noise in residential areas. If this limit were accepted, then from the data in Table 20.2, 9.7 M people or 17.8% of the population were subjected to unacceptable noise levels. Also a further 14.2 M or 25.6% of the population were within 5 dBA of the suggested limit, which some experts consider to be too high. It must be true that the position has deteriorated since 1970 because of the further increase of traffic volume and road construction. The Royal Commission on Environmental Pollution Fourth Report 1974 predicted that by 1980, 50% of the UK population or about 29 M people would be subjected to unacceptable traffic noise. It is true that road traffic noise does not cause damage to the physical ear mechanism, but it has other insidious effects. Continual exposure to this noise may speed up age-induced deafness or presbycousis, and it does disturb sleep quality and may lead to insomnia. At the very least, traffic noise disrupts relaxation and enjoyment in the home. Conversation, and the social pleasure derived from listening to radio, hi-fi recordings, and television becomes increasingly difficult as outside noise disturbance increases.

Government has made some recognition of the deterioration in the social environment by the Noise Insulation Act 1973, which empowers Local Authorities to make grants to householders. These grants can be used to offset the cost of sound insulation involving the installation of double glazing and special ventilation. Claims for grants are considered if the premises are adjacent to new improved roads brought into use since October 1972, and if the noise level is over L_{10} 68 dBA averaged over the hours of 0600 to 2400. There is also the possibility of discretionary grants for the occupiers of premises adjacent to new roads brought into use after October 1969. There are several ways of producing noise reduction. The most positive method is to reduce the noise at its source by the design and production of quieter engines and vehicles. It is certainly possible to reduce vehicle noise levels, but what are the desirable levels for particular types of vehicle? A Noise Advisory Council recommendation in 1974 was a progressive reduction of noise levels to 84 dBA by 1976. The Council also attempted to estimate the effect of reduced noise levels upon the population (Table 20.3). The calculated figures include an assumption that there will be a proportional increase in the vehicles in use and the population in the future.

Table 20.3.   Effect of reducing vehicle noise levels

Vehicle noise level assumptions	Population exposed to	
	L_{10} of 70 dBA or more	L_{10} of 65 dBA or more
1970 levels	14M	29M
5 dBA increase	4.3M	14M
10 dBA increase	0.9M	4.3M

Over a decade ago in 1963, Ernest Marples, the UK Minister of Transport, proposed that a maximum limit of 86 dBA be imposed for heavy vehicles. The last government proposed that the same limit should operate from 1980, and that there should be a reduction to 82 dBA for heavy vehicles by 1982. Progress towards the reduction of noise levels in the UK is appallingly slow. All sorts of reasons, including economic, EEC directives, and the sharp increases in fuel prices, have been put forward as excuses for not taking more positive action. There are modern types of heavy vehicles in use today with noise levels of less than 89 dBA. For example there is a Volvo 290 b.h.p. truck at 86 dBA, a Ford Transcontinental 340 b.h.p. truck at 88 dBA, and a British Leyland 200 b.h.p. truck at 86 to 88 dBA. Some authorities have stated that there are no practical difficulties to prevent the reduction of car noise to 80 dBA, or even to 75 dBA. However a quieter vehicle will always cost more, and research and development needs to be carried out by all vehicle manufacturers to prevent uncompetitive prices. The Motor Industries Research Association Transport and Road Research Laboratory has a government financed research programme in hand to produce two quieter engines. The objectives are to develop a new 200 b.h.p. and a 350 b.h.p. engine with a noise level not greater than 82 dBA, which is a level at present exceeded by many cars.

What is needed in the future are quieter vehicles as well as quieter engines. Vehicles at present in use could be less noisy if existing controls were more vigorously implemented. The M. O. T. test for vehicles over 3 years old now includes an examination of the exhaust system, but the vehicle is not required to be examined for any other noise source. The police force should be trained and required to carry out spot checks for noise levels at irregular intervals, especially in urban areas. If the police were as keen to check noise as they are to check speed limits, they would be providing a real social service to urban dwellers. Manufacturers should be required to quote noise levels as part of the specification of all new vehicles. This should apply to the whole range of vehicles, ranging from mopeds to the heaviest goods lorries. If low petrol or oil consumption and low noise levels were two of the main factors that determined vehicle purchase, then a considerable environmental advance would be achieved. The reduction of vehicle noise is an international problem, because at least 50% of all cars sold in the UK are manufactured abroad, including countries of the EEC. Also increasing numbers of heavy goods vehicles from the EEC enter and leave the UK each year. The EEC directives of 1970 and 1974 aimed to prevent the importation of vehicles into member countries with a noise level outside the range of 82 to 91 dBA according to their size. The concept of these directives is sound, but the full implementation by all member countries is apparently difficult to achieve. The reduction of vehicle noise in the UK therefore depends partly on international cooperation and action.

Road traffic noise could also be reduced by better control over traffic flow and volume. A reduction of heavy vehicle traffic on all roads would considerably reduce noise and some other environmental problems. At present, about 85% of UK goods, in terms of tonne miles, are transported by road. If some of the long haul traffic were transported by modern fast freight trains, there would be a considerable reduction in noise and road traffic congestion. This change would lower the volume of road traffic, and the need for additional motorways and by-passes would be reduced. Also if the EEC proposal to establish designated lorry routes restricted to certain roads were implemented, then this traffic and noise would be better distributed. Some Local Authorities are using traffic management schemes that aim to exclude heavy lorries from town centres and residential areas, at all times, or during prescribed hours. These schemes can only be operated if alternative routes exist, and where they will not cause undue traffic congestion. The alternative routing of traffic is sound environmental policy because it reduces noise, atmospheric pollution, and vibration damage to buildings, as well as helping road safety. Similarly, the limited construction of bypasses for trunk routes, and new motorways in rural areas has improved the quality of life for the inhabitants of small community towns and villages. County Council Authorities frequently do not give enough consideration to environmental effects when planning and developing new roads or improvements to existing roads. Planning officers should always have to assess traffic volume and noise volume as well as other factors. They should be obliged to ensure that the minimum number of people are exposed to all the hazards of road traffic noise. More attention should be given to the orientation of buildings with respect to main roads, and the use of landscape features to act as sound barriers and absorbers of traffic noise.

20.2 Aircraft Noise

Over the last decade there has been an increase in the noise nuisance from subsonic aircraft for several reasons. The noisier jet-engined aircraft have progressively replaced the earlier piston-engined and turbo-prop types, which were quieter, but still noisy. Increased numbers of civil aircraft have come into service, and consequently there has been a considerable increase in the number of air movements, both internally in the UK and also into and out of the major airports. The number of civil airports in the UK is not large, and the main noise disturbance is confined to a radius of about 16 km around them. But there are many people who work or live under the flight paths connecting airports, and the noise from aircraft passing overhead is inescapable. In 1977 there were about 436 000 flights into and out of the 26 main civil airports in the UK. In addition there are over 70 military airfields that cause considerable noise problems to the inhabitants of their vicinities.

Aircraft noise is variable and intermittent, and it is not continuous as in the case of road traffic noise. There are peak noise levels when aircraft are flying overhead, or are taking-off and landing at airports, and the peak frequency varies with the number and the type of aircraft, and the operational height. The noise is mainly produced from aircraft engines. Modern jet engine noise is a mixture of high frequencies from the compressor and turbine, and low frequencies produced from the rear jet exhaust nozzle. Attempts are being made to reduce the noise of existing engines in service, and these involve altering the shape of the exhaust nozzle by the fitting of retrofits or 'hush kits'. These can reduce noise levels by 3 to 5 PNdB on take-off, and 8 to 10 PNdB on landing, when fitted to Rolls-Royce Spey engines used in the Trident and BAC 111 aircraft. There is some reluctance by airline operators to fit hush kits, because they are expensive to buy, they increase fuel consumption, and reduce the aircraft operating range. An alternative method of reducing engine noise is to fit new fans, but the cost of this is higher than fitting hush kits. Economic costs are of increasing importance to airline operators, especially as the price of fuel continues to increase rapidly. A US cost benefit study in relation to DC8–50 aircraft was made in 1975 (Table 20.4). These costs must be considered in the context of a possible further operational life of about 7 years for the DC8–50 aircraft. In 1973, British Airways studied the cost of fitting hush kits to all the aircraft that required them. The estimated cost at that time was £100 M, which was about 4% of the annual revenue of the airline. Similarly, the cost of reducing jet engine noise on all the older types of aircraft in service all over the world would be very high. It has been estimated that 700 aircraft would require hush kits, or 50% of the civil aircraft that entered service prior to 1969. Also all these planes have an average regular service life of only 15 years, so they would all be withdrawn from operation by the mid 1980s. When all these facts are considered, it is not surprising that universal modification of the noisiest jet aircraft has not been carried out. In the UK, the Ministry of Aerospace announced in 1973 that hush kits must be fitted to all BAC 111 planes coming into service after January 1976. This was helpful in reducing noise, but it did not affect the immediate short term noise problem with this aircraft. Planes that came into service after 1969–70 have been fitted with the newer high bypass turbo-fan engines. A good example is the Lockheed TriStar

Table 20.4. US cost benefit study on aircraft noise reduction

PNdB noise reduction	Hush kit	New fans	New engines	Replaced by DC10–30
At take-off	5	10	15	13
On landing	12	15	17	15
Cost (dollars)	560 000	1.7M	6.2M	4.7M
Average cost per ticket (dollars)	1	4.5	16	2.5

Table 20.5. Aircraft noise levels. Column A, data from Consumer Association (CA) noise survey measured under flight paths at London Heathrow, Manchester Ringway, and Luton airports in 1974; at a distance of 1.6 km, or 4.8 km (Luton), from the take-off point. The figures are averages from 1700 recordings. Columns B and C, British Airports Authority (BAA) levels in 1971 based upon the total weight of the aircraft

Aircraft type	Entered service	Average survey noise level 1974 (PNdB)	BAA permit-ted take-off levels (PNdB)	BAA levels with hush kits fitted (PNdB)
		A	B	C
VC-10	1965	110.6		
DC-8	1967	112.4	114	106
Boeing 707	1967	115.4	114.5	107
Boeing 737	1967	110.7		
Boeing 727	1967	112.9	112	107
DC-9	1968	111.7	108	104.5
BAC 111	1968	119		
HS Trident 3	1971	119.5		
Boeing 747	1972	114.7	114	
DC-10	1971	110.6	100	
Lockheed Tristar 1093	1973	105.1		

airliner fitted with RB 211 engines, that have a noise level of 10–12 PNdB lower than the previous Rolls Royce type of engine. This new engine design reduces the internal gas velocity and mixing, and has a higher bypass ratio, resulting in reduced exhaust nozzle noise. The engine ducts are also lined with noise absorbent material. The comparative noise levels of the main aircraft at present in service are shown in Table 20.5.

The table includes seven types of aircraft that came into service before 1969, and they will be phased out in the early 1980s. In considering the noise levels of aircraft, it is relevant to remember that an increase of 5 PNdB doubles the noise level. Also that London Heathrow, Gatwick, and Ringway Manchester airports prescribe the maximum noise level for take-offs at 110 PNdB in daylight, and 102 PNdB during darkness. By comparison the limit for daylight at New York airport is 112 PNdB. Noise checks made by the Department of Trade during April, May, and June in 1975, showed that 70% of the 693 take-offs by Boeing 707, VC-10, Boeing 747, DC-8, BAC 111, and Trident aircraft infringed the 110 PNdB level at Heathrow airport. By far the noisiest aircraft was the Trident, and it was responsible for half the infringements. The quietest aircraft were the TriStar and Boeing 747, which are fitted with the new turbo-fan engines.

Helicopters are increasingly in operational use, and they are also noisy aircraft. Most people do not consider that they are a particular hazard at present, but they are being used more widely by internal airlines, the armed forces, private companies and individuals. At present there is no legislation to control noise levels when operating from private land, but there are designated authorized helicopter routes over central London. Noise level for helicopter take-offs is prescribed as 92 to 97 PNdB, which indicates that they are only marginally less noisy than the quieter airliners.

Public protest and concern about aircraft noise markedly increased from 1974 onwards. This was partly caused by the trials and proving flights of the BAC/Sud Aviation Concorde aircraft, and its entry into service in 1976. The lengthy design and development period of 12 years meant that the aircraft was not fitted with the quieter turbo-fan engines, which were not available until the late 1960's. Therefore considerable noise during take-off and landing of Concorde was inherent in its operation, and to this was added the sonic boom during flight. This is an effect produced by all civil and military supersonic aircraft, when they pass from subsonic to supersonic speeds in flight, or they travel faster than the speed of sound. At subsonic speeds all aircraft produce forward air pressure waves, but at supersonic speeds they outrun the forward waves. This produces violent air pressure changes that move out from behind the aircraft as a multiple shock wave, causing the so-called sonic boom. This is not a single occurrence, but occurs continuously as a carpet 80 to 128 km wide, so long as supersonic speeds are maintained. The boom or bang may be heard as a rumble or a sharp crack, and ground observers usually hear it without any warning from aircraft engine noise. The boom pressure waves can cause psychological disturbance and damage to buildings, but careful control of aircraft flight has virtually eliminated boom nuisance over land. Concorde only reaches supersonic speed at a height of about 13 000 to 16 400 m, by which time it has travelled 160 km from the airport. Since many airports are within 160 km of the sea, Concorde's flight over land is subsonic, and supersonic speeds only take place over the sea, which causes the minimum of noise disturbance. Where this situation does not apply, then flight restrictions may be imposed, to ensure that the aircraft's speed is subsonic over heavily populated areas. However there is still the problem of Concorde's engine noise. The aircraft manufacturers estimated the noise level at 105 to 108 PNdB at take off, but some observers have recorded noise levels up to 120 PNdB, which is twice as loud as the noise level produced by a DC-10 or Boeing 737 aircraft. Operational experience of Concorde has shown that the noise level at take-off can be kept within the 110 to 112 PNdB airport limits in the UK and the US. Airline operators accept the noisiness of Concorde, but press the advantages of a considerable reduction of supersonic flying time against some possible infringement of maximum noise limits. There now appears to be general acceptance of Concorde flights, wherever these operate around the world. This shows how flight handling and strict operating procedures can produce a reduction in aircraft noise. Future operation of more supersonic airliners by the UK, and probably the US and USSR, alongside quieter subsonic aircraft, may alter their acceptance by people living and working in the vicinity of major airports.

The NNI for aircraft noise expresses the maximum noise produced by aircraft flying overhead, and the degree of the annoyance that is caused. The Wilson Report recommended that noise levels over 35 NNI caused annoyance, and over 50 NNI created intolerable annoyance. The report also recommended that no new major housing development should be sited in noise zones of 40 NNI or more. In 1973, it was estimated that 5 M residents were affected by aircraft noise of over 35 NNI in the whole of the UK. The Noise Advisory Council estimated that in 1970–71, the number of residents subjected to noise of over 35 NNI was 2.3 M around London Heathrow; and 300 000 around Manchester Ringway, Gatwick and Luton airports. At Heathrow, 2.2 M people live in the 35 to 45 NNI zone, and 300 000 in the over 45 NNI zone. The Consumer Association survey 1974 estimated that at Heathrow the NNI value ranged from 76 to 77 at 1.6 km from the runway, at Ringway it was 61 to 65 NNI at the same distance and at Luton the NNI was 50 to 55 at 4.8 km from the runway. Another survey sampled residents' opinion of noise in 1973. It was found that a majority of people accepted a noise level of 93 to 105 PNdB without complaint, but 69% were severely disturbed by noise levels of over 103 PNdB. Levels of over 105 PNdB were generally considered to be completely unacceptable. Considering these levels in relation to aircraft noise, as shown in Figure 20.2, then it seems that only the noise level of the Lockheed TriStar is acceptable for residents in close proximity to Heathrow. All these surveys have apparently been concerned with localities close to airports, but the residents who live beneath aircraft flight paths must not be overlooked. They are experiencing increasing noise nuisance as the number of daily flights grow each year, especially in the summer months, when the social amenity of gardens and outdoor sports and events are enjoyed.

The Government and Airport Authorities are not entirely oblivious to the nuisance of aircraft noise. The British Airports Authority have regulations made under the Airports Authority Act 1965 for the approach and take-off of aircraft. The height and angle of landing approach is prescribed, and also the angle of take-off and engine power reduction at 305 metres, to reduce the duration of maximum engine power output. Approach and take-off flight paths are carefully chosen to avoid densely populated areas. There are also restrictions placed upon the number of landings and take-offs during the hours of darkness, to reduce the

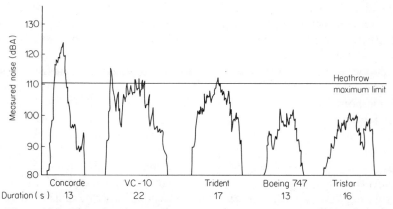

Figure 20.2. Aircraft noise measurements at Hatton Green, Middlesex, on 7 July 1975.

interference caused to rest and sleep. London Airport at Heathrow has prescribed a maximum noise level of 98 dBA in daylight and 90 dBA in darkness for the overflying of residential areas. The Air Navigation (Noise Certification) Act 1970 specifies that British and foreign subsonic jet aircraft must have clearance certificates that comply with noise limits, in order to operate in and out of British airports. Despite these efforts to control aircraft noise many infringements continually occur. This is because the implementation of the requirements depends upon the aircraft pilots, and there are numerous reasons for their aberrant behaviour, including the weather and aircraft safety. The British Airports Authority Act 1965 allows grants to be given to householders towards the cost of sound insulation and ventilation of their premises. If the noise levels are 40 to 55 NNI, scaled grants are given. Research at Heriot-Watt University, Edinburgh, has shown that adequate sound insulation can reduce the internal noise in rooms by a mean value of 13 to 14 dBA, which means a noise reduction of over 50%. Another indication of the Government attitude to airport noise and its effects was the decision in 1973 to ignore the Roskill Commission Report recommendations, and to propose that a third London airport be sited at Maplin. This site was preferred to several other more economic alternatives largely for environmental and noise disturbance reasons. Despite all these efforts to control and reduce aircraft noise nuisance, it is still a serious problem for millions of people.

Measures to reduce the level of aircraft noise must be mainly directed towards aircraft engines, assuming it is quite impracticable and unacceptable to reduce the number of aircraft scheduled flights. Proposals to reduce jet-engine noise are either short term or long term. Short term measures apply to aircraft at present in service. These were designed for a specific engine size and power output. Therefore it is usually impossible or uneconomic to replace the originally specified engine with a quieter type. However certain measures can be taken to reduce aircraft engine noise. Hush kits could be compulsorily fitted to specified types of very noisy aircraft, for example Trident, BAC 111, and Boeing 707. The alternative is to remove these planes from operational service. Because of economic considerations, it is reasonable to require hush kits to be fitted to those aircraft that have a remaining operational life of 7 years. To cover the cost it would probably be necessary to increase air fares, but this should be socially acceptable. Passengers who enjoy the advantages of modern air travel should contribute towards the quality of life of people who are socially disadvantaged through noise.

All airport authorities should be obliged to provide muffler bays to reduce the ground noise caused when engines are being tested, or warmed up before take-off. There should be frequent reviews of the number of night flights into and out of British airports, and the type of aircraft that operate them. Night flights should be restricted only to the quieter types of aircraft. All airport authorities should have automatic noise measuring equipment to monitor individual aircraft noise, and pilot performance on take-off and landing. Manchester Ringway airport instituted a scheme to encourage quieter take-offs in 1974. Airlines receive a rebate on their landing charges for all take-offs that are within the Ringway noise limits. For example, if hush kits are fitted to a BAC 111 at a cost of £103 000, the cost is recouped after 8500 quiet take-offs. One airline has fitted hush kits to its Boeing 737s, and using a modified take-off technique the noise level has been reduced to 98 PNdB, and the airline has qualified for a 98% bonus over a 1 year period. The total effect of the Ringway scheme has been a reduction of take-off noise level by an average of 3 PNdB, and complaints from nearby residents have fallen sharply. Frankfurt airport also operates a points bonus scheme to encourage noise reduction. It is therefore possible to reduce take-off noise even for older aircraft, but improved pilot take-off techniques are necessary, and these require to be monitored. Lastly the grant position for sound insulation should also be reviewed. The basis of grants should be reconsidered, and 35 NNI should be adopted as qualifying for percentage grant and 50 NNI for a 100% grant. If these measures were implemented, then more people would be encouraged to carry out sound insulation, because the proportional cost to each householder would be reduced. All these suggested changes would help to reduce the disturbing and psychological effects of noise upon sleep, and improve the health of people near airports, and beneath flight paths.

Long term noise reduction measures are concerned with the future development of aircraft and airports. Aircraft and engine manufacturers should receive every incentive to design and produce new and quieter power units and airplanes. Control over this is available through the Noise Certification Scheme 1971, and the Air Navigation (Noise Certification) order 1970. These regulations are in accordance with the ICAO 1971 recommendations, which should be widely implemented by all major aircraft manufacturing companies. Future aircraft in the 1980s and 1990s will no doubt be larger and more powerful, but they must also be much quieter. In conjunction with the above controls on manufacture, the airport authorities should reduce their maximum permissible noise levels from the present 110–112 PNdB in daylight, and take rigorous steps to enforce the newer noise levels. The siting or extension of new or existing airport facilities must also be much more carefully considered than in the past. The Roskill enquiry was an example of how such proposals can be dealt with. There must be a mechanism to allow public views to be expressed, and allow the planners and technical experts to be publicly questioned. In future it should no longer be possible for thousands of people to be subjected to excessive aircraft noise. Those already disadvantaged should have their environment improved, even if this means restricting the size of present airports, and providing new ones away from large conurbations.

20.3 Rail Traffic Noise

The noise from trains is not generally regarded as a serious nuisance by the majority of people. This may be because the noise is generally of a lower frequency than that of road vehicles, and also most railway track runs through rural areas. But where buildings are sited beside railway tracks, and especially where engine testing and shunting is carried out, there is noise disturbance.

Railway noise in the UK has generally decreased as a result of several developments. By 1967, the effect of the infamous 'Beeching Policy' was to reduce drastically the length of trunk routes by nearly 50%, the number of passenger trains operating by 84%, and freight trains by 25%. Indeed trains disappeared altogether from large areas of the rural countryside. Marshalling yards were reduced by 90%, and goods yards by 67%, and these cuts greatly reduced the noise nuisance from freight shunting operations, especially at night. New types of freight rolling stock have been designed that are quieter to load and have automatic couplings, so reducing noise in marshalling yards. The introduction of liner trains using containers has greatly reduced the amount of shunting and the use of intermediate yards. The last steam locomotives were withdrawn from service in 1968, and they were replaced by diesel–electric traction, which is much less noisy, especially when starting and accelerating. British Rail research and development during the 1970s has produced improved types of locomotives and new designs for passenger coaches. The high speed train or Inter-city 125 was introduced into service in October 1976, and work is in progress on an advanced passenger train for introduction in the 1980s. Passenger comfort has been considerably improved particularly in relation to noise. The increasing use of welded track, and improved coach suspension and air-conditioned coaches on the 125 Inter-city trains, have all contributed to passenger noise reduction. Parallel with these developments has been a further electrification of some routes, which use all-electric locomotives that produce no atmospheric pollution and comparatively little noise.

The railways appear to be a good example of how the developments in modern technology and design can contribute to a reduction of noise. There has never been any specific noise legislation for rail traffic, and it was outside the scope of the Noise Abatement Act 1960. The Noise Advisory Council Report on 'Neighbourhood Noise' 1971, noted the non-inclusion of rail traffic within noise legislation, and did not propose any change in this position. This is still true today.

CHAPTER 21

Occupational Noise

This type of noise, together with road traffic noise, are the two major sources of noise that affect millions of people. Broadly, occupational noise is largely produced by industrial machines and processes. But also included in this category is the increasing noise produced in the home from washing machines, spin dryers, food mixers, sink waste grinders and vacuum cleaners. (Table 21.1). In industry, workers are subjected to noise for periods of up to 8 hours per day, 5 days per week, 48 weeks in the year. Some industrial processes are much noisier than others, but it is known that permanent hearing damage is often found in connection with weaving, ship-building, boiler-making, forging, pressing, and blasting operations.

Industrial noise is complex and varies with the design, direction of movement of working parts, and the method of mounting of machines. The noise is often produced in three stages. There is an initial disturbance at the point of origin of the sound; followed by amplification, often caused by the resonance of the machine parts, the workpiece or the floor; and finally radiation of the sound to the surrounding environment. The specific noise produced by any one machine is often a combination of these three stages. Industrial noise can be classified into four types. Impact and percussive noise is produced by presses, punch and stamp machines, pneumatic drills, milling machines, cutters and routers. The impact noise is caused when two surfaces meet each other, sometimes at high speed, and vibration occurs at the point of contact followed by amplification and resonance. Sustained mechanical vibrations are caused by random movements between two surfaces in contact. The movements can be in six different directions i.e. upwards and downwards, side to side, forwards and backwards, yawing and rotary movement about a horizontal axis. The total vibration of any one machine tool is usually a combination of these directional movements e.g. drilling, turning, and milling. The third type of industrial noise is aerodynamic noise. This differs from the first and second because no metallic vibration or resonance is involved. Aerodynamic noise is produced by a blow lamp or torch, fans, and dust extractors. There are different types of aerodynamic noise, but generally all involve mixing and, or expansion of a fast moving stream of air or gas, movement past or across a solid structure, and the production of air disturbance, or vortex effects. Lastly, electromagnetic vibration can occur in electric motors, alternators, and transformers. The vibration is caused by the forces produced

Table 21.1. Occupational noise levels

Level (dB)	Industrial		Domestic
130	Steel plate riveting		
126	Oxygen torch		
122	Pneumatic metal chipper		
120	Boiler makers' shop		
112	Textile loom, fettling shop		
110	Circular saw		
105	Pile driver at 15 m		
103	Farm tractor	103	Powered lawn mower
101	Newspaper press		
100	Coal face drill		
95	Bench lathe, some computers		
90	Milling machine	90	Food blender
86	Bed press		
85	High speed drill	85	Alarm clock at 1 m
		83	Waste grinder
82	Key press machine	82	Washing machine
		76	Dish washer
		69	Vacuum cleaner at 3 m
65	Typing pool, canteen (dBA)	65	Toilet flush
60	Supermarket (dBA)	60	Conversation at 1 m
42	Quiet office (dBA)		
40	Public library		
34	Soft whishper at 1.5 m (dBA)		
		32	London flat at 24.00
20	Church		
10	Normal breathing		
0	Threshold of hearing		

between the armature and the field magnet, or the rotor and stator. In transformers, the vibrations can result from the deformation of the core caused by magnetostriction. This type of noise is often the AC current frequency with additional harmonics. Where rotary movement is involved, the total noise includes an aerodynamic component as well as the elctromagnetic vibration.

The combination of these different categories of noise can be appreciated by considering the internal combustion engine. In the cylinders there is fuel combustion and rapid gas expansion. Energy is produced in the form of pressure, heat, and noise. Percussive combustion noise causes vibrations in the cylinder block, crank case, timing case, and valve cover, where resonance and multiple harmonics are produced. The combustion explosion acts on the pistons, and their motion activates the crankshaft, valves, and flywheel. This movement gives rise to mechanical vibration noise. The opening of the inlet valves admits fuel mixture from the induction manifold to the cylinder, and the opening of the exhaust valves allows the exit of gases into the exhaust manifold. This rapid movement of gases causes aerodynamic noise as they pass into the exhaust system, and the noise is further amplified and resonates in the exhaust pipe and silencers. Further aerodynamic noise comes from the cooling fan, and this will increase to about 60 dB by a ten-fold increase of engine speed. The noise level of the whole engine is affected by the engine speed and cylinder capacity. A ten-fold increase of engine speed raises the noise level by 50 dB, but a ten-fold increase in engine capacity only increases the noise level by 17 dB. Therefore for the same power output, a small capacity, fast revving engine is noisier than a larger, slow revving engine.

Noisy industrial processes and conditions produce hearing loss to the workers involved. Detailed occupational noise studies by a team from Manchester University in 1971 confirmed this (Table 21.2).

It is regrettably true, at least since the nineteenth century, that many industrial workers have suffered hearing loss. Some quoted statistics show the extent of this over the whole range of industry. In the UK the precise number of workers who have hearing loss is probably not known. R. T. Taylor estimated that there were probably 5000 companies with employees at risk, in a working population of 25.2 M in 1975. In 1962, an Industrial Welfare Society survey found that 24% of workers in 55 firms were working in such noisy conditions that the employees were auditorily isolated, and unable to communicate with each other. In the various industrial surveys that have been made the number of isolated workers varied from 90% in a shipbuilding and repairing firm, to 75% in a textile firm, and 50% in drop-forging and printing firms. In 1971, the National Coal Board tested 394 miners, and found that 50% had appreciable hearing loss. A survey by the Department of Employment in 1973 suggested that at least 600 000 people work indoors in factories, where the conditions could give rise to noise-induced hearing loss. Another estimate stated that 10–14% of UK workers suffer some

Table 21.2. Hearing loss by workers in specific jobs*

Occupation	Worker's age (years)	Length of occupational service (years)	Hearing loss (dB)
Gravure printers	55.3	21.8	35
Iron moulders	54.8	36.4	43
Bus drivers	53.7	21.2	44
Cardroom operatives	59.8	30.5	52
Boiler makers	55.2	35.5	55
Weavers	55.0	31.5	58
Fettlers (casting, metal chipping)	53.5	21.3	65

*The average normal hearing loss for people in the 50 to 60 age group is 30 dB.

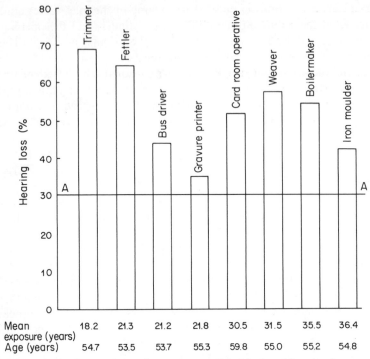

Figure 21.1. Comparison of some occupational-induced hearing losses.
Line A–A, normal hearing loss in the 50 to 60 age group.

loss of hearing as a result of their occupational environment. Yet another researcher estimated that about 1M, or 4% of employees, work in conditions which will make them partially deaf after 20 years' service (see Figure 21.1) It should be recognized that the information available about occupational hearing loss is not always very precise. The age and length of service of the various categories of workers, the noise levels they experience, the daily duration, the amount of hearing damage, and the extent to which ear protection is available and used, are all important items of information that need to be known, in order to accurately assess industrial situations. It is also apparent that not enough medical knowledge is available about the precise effects of occupational noise upon individuals. This may be because at present, very few individuals have been awarded compensation for hearing loss. However, despite these short comings, there is enough information to show that there is an unsatisfactory situation in industry at present. This position may be improved when the Health and Safety at Work Act 1974 becomes fully implemented, together with codes of practice incorporating specified industrial maximum noise levels.

The Department of Employment produced a 'Code of Practice for Reducing the Exposure of Employed Persons to Noise' in 1972. This proposed that 90 dBA should be the maximum steady noise level for workers over a continuous 8 hour period. The noise level and time of exposure varies very much in different

Table 21.3. Noise limits and daily exposure

Duration	Noise limit (dBA)—daily exposure
8 hours	90
4 "	93
2 "	96
1 "	99
30 minutes	102
15 "	105
7 "	108
4 "	111
2 "	114
1 "	117
30 seconds	120

industries, and the detailed code of practice recommendations are given in Table 21.3. These limits, if applied, mean that 90% of the working population would suffer a permanent hearing loss of less than 20 dB after 50 years of exposure. This degree of hearing loss is not very serious, even when added to the natural hearing deterioration through age. Some industrial hearing experts consider that the 8 hour limit should be reduced to 80 dBA, or 85 dBA to provide a greater margin of safety and protection for all workers. A maximum noise level of 80 dBA would increase the population figure to 96% instead of 90%. At present there is no statutory Act which specifies an industrial noise limit, but the code of practice limits are used by Health and Safety Inspectors.

Additional recommended noise levels have also been proposed, and these have been given in Table 19.2. It is evident from Figure 21.1 that there are many industrial occupations where workers are at severe risk from hearing damage. This can be reduced by the use of some form of ear protection. There are three chief types of protection available for use in industry. The crudest and simplest is disposable wads or ear plugs of flexible material. The best type available at present is glass down, which is more efficient than cotton wool, and has not got the irritant properties of glass fibre. The other type of internal protection is provided by moulded ear plugs. These must be of the correct size, and comfortable to wear, but need regular disinfecting. One type has a cut-off valve that closes when the external noise level reaches a preset limit. Ear plugs can reduce noise levels by 15 to 35 dB according to their design and fit. The alternative to ear plugs is externally worn ear defenders or muffs. These are usually more comfortable and easier to wear, and can reduce noise levels by 35 to 45 dB, depending upon the frequency of the received sound. One type of ear muff is now available that has an electronic facility to cut out loud noise, but allows the wearer to hear quiet sounds and speech. Ear muffs are more expensive than ear plugs but have wearer advantages, and are more efficient ear protectors. The wearing of ear protectors has obvious advantages, but some disadvantages. Their use is resisted by many workers for several reasons. Many types are not comfortable to wear, and isolate the wearer from Communication with fellow workers and the background noises of the work situation. Miners, for example, complain that ear protectors are a safety hazard, because they cannot hear sounds of potential roof falls, and verbal warnings. Older workers, who have never been accustomed to wear ear protection, see no need for them, because they have always accepted that the noise goes with the job. There is also the attitude that it is effeminate to wear ear protectors, and strong men do not need them. The problems involved in the use of ear protection are therefore physical and psychological. Education of all workers is required in respect of the long term benefits to the individual, as well as firmly applied company regulations and the support of the trade unions. Obviously ear protection is not the only answer to the problem of industrial noise.

The record of compensation awarded for occupational hearing loss, and hence legal recognition of this, is very meagre. The former Industrial Injuries Act only accepted injury from a specific industrial accident as qualifying for compensation. In 1971, Frank Berry made legal history by being awarded £1250 compensation for hearing deterioration experienced as a fettler, involving chipping and trimming metal off propeller castings in shop conditions of 115 to 120 dB. The Industrial Injuries Act 1974 consequently recognized injury in the metal manufacturing, ship building and repairing industries, as elegible for compensation. In 1976, the first textile worker from Lancashire received £2000 for industrial deafness injury. More successful compensation award cases could help to improve considerably the working conditions of the noisy industries.

An interesting noise contradiction is the use of so-called background white noise in commercial offices. This type of noise may be soothing ocean wave sounds, or comparable electronic sounds purposely produced by a white noise generator. It is being used to preserve the confidentiality of conversations in open-plan offices, and for insomnia treatment, and another suggested use is in houses to blanket out excessive neighbourhood noise.

Occupational noise is present in the working environment of many thousands of people for up to 1900 hours per year, and for 49 years of their lives. The benefits of noise reduction are very worth while considering. Thousands of people would no longer suffer progressive hearing damage, verbal communication between workers would be enhanced, improved efficiency and a reduction of accidents could result, as well as a general all round improvement in working conditions.

There are a number of ways in which industrial noise can be reduced. Ideally noise output should be reduced at its source, and much can be achieved by a careful consideration of the mounting and location of machinery and equipment, when it is first installed. Anti-vibration mountings of various types are available, and gaskets for pipe and duct connections, and braided connectors can all reduce noise. Some noisy machines and processes can be isolated or surrounded by baffles, in an attempt to restrict a noisy work area and prevent noise propagation over the entire working area. The reduction of noise is a technical and skilled operation, and it is often necessary to employ industrial auditory consultants to achieve efficient results. Their advice should also be sought when planning new buildings and extensions, and deciding upon the layout and location of machinery. More effort should be made to reduce the noise output from machinery. This involves a definite commitment by designers and manufacturers to produce less noisy industrial equipment and machinery. Design specifications that state loading, power, work rate, and weight should always include noise levels at different working speeds. In the long term it is often less costly to design and produce a quieter new machine, than to attempt to reduce the noise of an existing one, often with very varying success.

It may be that a reduction of noise at source is not possible, or attempted reduction is unsuccessful. In this case, if the noise levels are approaching 90 dBA, then ear protection of some type should be provided and worn by all employees. The provision of ear protectors may be required by the Health and Safety Inspectorate. But management and supervisory staff have a duty to ensure that ear protectors are worn as necessary, to provide protection from peak noise levels. Industrial establishments with a noisy environment should regularly monitor noise levels and make the results available to their employees, so that they are aware of the risks at work. It is very desirable that periodic audiograms are taken for all employees at risk to identify threshold shifts. The equipment for these is relatively expensive, but audiogram checks can be carried out by specialist firms on a regular contract basis.

Apart from noise reduction measures in places of work, improvements can be made in the requirements for noise limits. Noise reduction can only be seen to be achieved when compared with specific maximum noise levels. The defining of noise levels is a technical matter and the determination of general noise levels is too crude to be acceptable. The component frequencies of noise and the peak levels produced are very important. Often it is the periodic exposure to peak hazard levels that causes hearing damage, even though the general noise level may be below 90 dBA. Codes of practice for specifically noisy machinery and operations should always be available, and it is to be hoped that these will be produced during the progressive operation of the Health and Safety at Work Act.

There needs to be a significant improvement in the noise compensation situation. The number of compensation awards granted bears no relationship to the number of workers who have suffered occupational ear damage. If more compensation had to be paid by employers, then it is very likely that industrial noise would receive much more serious attention in certain industries. It is quite iniquitous that external body injuries are quite acceptable for compensation purposes, but monetary awards are rarely given for hearing damage.

More research is required into almost all aspects of industrial noise. There is insufficient information about the long term effects of noise upon individual workers, the component frequencies of specific operation noises and their effects, the efficient methods of reducing noise, and the design requirements of quieter machines and processes.

CHAPTER 22

Neighbourhood Noise

This has been defined by the Noise Advisory Council as 'the variety of sources of noise which may cause disturbance and annoyance to the general public. It is an unjustifiable interference with ordinary human comfort and well-being'. Neighbourhood noise includes disturbance that may be present in the home, and in public places of entertainment or in the open air, but excludes industrial, aircraft, and road traffic noise (see Table 22.1).

The total extent of the nuisance from neighbourhood noise in the UK is not known. The scanty information available is usually derived from public complaints, and many people are annoyed or disturbed by noise, but do not officially complain. In 1969, the Ministry of Housing and Local Government surveyed Local Authorities, and of those that reported, 39% stated that they had a serious industrial noise situation, 16% had a slight problem, and 36% had no problem. Overall, 76% of the London and County Boroughs had a serious industrial problem, as against 25% of other Local Authorities. The survey was based upon complaints received over the two year period 1967 to 1969. The Association of Public Health Inspectors also collated information, and found that noise nuisance complaints had increased by nearly 25% between 1966 and 1969. No doubt complaints have continued to increase since this period because the environment appears to be becoming more noisy.

Noise in the home impinges upon domestic life, and the sources may be the antisocial activities of neighbours in respect of loud TV and radio sets, parties, children playing, dogs, motor cars, and motor cycles, and also noise from industrial premises where houses are situated nearby. Much of the noise caused by neighbours is tolerated, or dealt with by personal complaints to the offender. Only in a minority of incidents are complaints made to a Local Authority. Noise from factories occurs particularly in urban areas, and as shown above provokes a relatively large number of complaints to Local Authorities. This type of neighbourhood noise may be continuous, or transient depending upon the weather and time of day.

Sound travels through air, and its intensity is inversely proportional to the square of the distance from the source. Alternatively it can be stated that if the distance from the source is doubled, the noise intensity is reduced by 6 dB. The distance between a house and an industrial building therefore affects the noise level received. Noise may also appear to be directional to some recipients, and

216

Table 22.1.   Neighbourhood noise levels

Level (dB)	
200	Saturn rocket lift-off at 300 m
140	Carrier jet plane take-off
120	Supersonic plane take-off, some pop groups
115	Jumbo-jet plane take-off
110	Jet plane flying overhead
108	Pneumatic drill at 1 m
104	Drophammer pile driver at 15 m
100	Hovercraft at 18 m, some discotheques
96	Motor cycle at 8 m
94	Bulldozer at 15 m, twin-cylinder motor cycle
93	Tractor-scraper at 15 m, heavy truck at 15 m
92	Heavy diesel lorry at 7 m
90	Train whistle at 150 m, sports car
85	Train at 30 m
82	Tree saw at 15 m
80	Earth digger at 15 m, small car, motor scooter
77	Underground train ('tube')
76	Mechanical shovel at 15 m
75	Dumper, concrete mixer at 15 m
64	Concrete vibrator at 15 m
25	Still day in a country field (dBA)

this is caused by changing wind direction and the daily air temperature gradient. Warm air transmits sound more rapidly than cooler air. During the day, the sun and fuel combustion warms the air, but at night the warmer air rises and the air nearest to the ground is cooler, so creating a temperature gradient. This partly explains why some sounds appear to carry more at night than in the day, but low wind velocity and the relative absence of background noise at night also helps.

It is significant that one of the commonest complaints about industrial neighbourhood noise is that it disturbs or prevents sleep. People vary in their tolerance or intolerance of noise at night. Steady noise of medium to low frequencies is usually acceptable, but impulsive or percussive noises and those of high frequencies are often unacceptable. Some experts have advocated that for 80% of the population, the maximum neighbourhood noise level should be a peak of 45–50 dBA against a background noise level of 35 dBA. When complaints are made, the onus to reduce noise is upon the industrial establishment, under the statute of the Control of Pollution Act, 1974. Many of the noisy neighbourhood noise situations are the direct result of bad site planning. Local Authority

Planning Departments are responsible for approving the siting of new industrial premises or new housing development. Often in the past, and even today, the environmental effects upon buildings have not been sufficiently explored. Prevailing wind directions, and the reflection of sound waves from terrain and existing buildings, affect sound propagation. The type of industrial building, its sound absorption factor, the noise characteristics of the industrial processes, and hours of working are other factors that need to be considered. Outside the industrial premises, the effect of the walls and landscape features upon sound reflection and/or absorption are only just beginning to be studied and implemented.

Noise in open-air spaces is usually a transient or periodic nuisance depending upon the various activities that are taking place. Noise from a football ground can reach a level of 110 dB when the home team score a goal, but the noise only occurs for 90 or 180 minutes per week when a match is being played, and football crowd attendances are generally diminishing. Motor sports meetings of various types, athletics meetings, fêtes, carnivals, and open-air pop concerts may all be sources of periodic noise in local neighbourhoods. The increasing use of public address equipment and amplified music can cause complaints, especially when pop festivals lasting many hours may produce peak noise levels of up to 120 dB, at a distance of 90 metres from the stage. Means of control of open air noise are available through bye-laws and regulations implemented by Local Authorities and various authorized sports bodies. These are described in detail in Chapter 28.

Indoor places of entertainment such as dance halls, licensed premises, and discotheques can cause noise nuisance to nearby residents, especially late in the evening. Discotheques may be a potential noise hazard to the dancers, because the equipment may produce noise levels of 90 to 110 dBA at varying points around the room. Research carried out at Leeds University showed that even spasmodic visits to discotheques can cause an auditory temporary threshold shift in teenagers, which requires days for complete recovery. Another survey concluded that one in five school leavers or apprentices have some hearing loss. A recommended maximum noise level of 96 dB has been proposed for discotheques, and in 1974 Leeds City Council restricted maximum amplifier noise to 93 dBA. Limited noise levels can be controlled by an electronic cut-off in the amplifiers that operates when the pre-set noise level is exceeded for more than 6 seconds. Control over indoor noise can be implemented by Local Authorities or by licensing magistrates, when granting entertainment licences.

A growing number of complaints arise from the activities of the construction industry, for example demolition, building, and road construction work. Noise nuisance is caused to many people particularly where the working site is located near houses, blocks of flats, commercial and retail premises in urban areas. Construction operations are now highly mechanized and require the use of heavy and noisy plant and equipment. Table 22.1 shows that noise levels can vary from 64 dB for a concrete vibrator, to 108 dB for an unmuffled pneumatic drill. There are means available to reduce the noise output of some construction machinery, for instance mufflers can reduce pneumatic road drill noise by 10 dBA, and

Table 22.2. Indoor noise standards (L_{10} in dBA)

Location	Day	Night
School classrooms	45	—
General offices	55–60	—
Private offices	45–50	—
Lecture theatres, conference rooms	30	—
Suburban areas away from main traffic routes	45	35
Busy urban areas	50	35
Rural areas	40	30

hydraulic pile drivers can be used instead of the very noisy drop hammer pile drivers. Control over construction site noise can be exercised by Local Authorities under Part 3 of the Control of Pollution Act 1974.

There are several ways in which an improvement in neighbourhood noise could be achieved. The Control of Pollution Act 1974 gives Local Authorities powers to impose requirements concerning the way in which construction works are to be carried out, the plant and machinery to be used, the hours of working, and the noise levels that may be emitted. If these powers were enforced, construction contractors would have to reduce the noise levels of pneumatic drills, compressors, dumpers, and diesel generators. The environment of residential and commercial premises dwellers would be further improved by better noise control in places of entertainment. Licences for music and dancing are granted by local magistrates supported by Local Authorities. Local justices should prescribe maximum permissible noise levels when granting all initial licences and when renewing them, and so assist noise abatement. There is a need to improve the control of noise in open air spaces. Noise in these areas can be controlled by bye-laws which prescribe maximum noise levels for loudspeakers, open air concerts, radios, model aircraft, acoustic bird scarers, etc. The existence and operation of bye-laws for this type of noise is very variable in different local areas. More could be done by LAs, and the DOE should supply detailed guidelines to LAs in respect of noise levels and their measurement.

The general attitude to noise by the various controlling authorities should be improved. The Noise Advisory Council in a report on neighbourhood noise in 1971, proposed the establishment of noise abatement zones. Powers to establish these zones are included in the Control of Pollution Act 1974. The concept aims at reducing ambient noise levels in residential areas, and allows control of noise output from specific industrial premises. A guide for ambient noise levels was put forward in the Wilson Report in 1963. (Table 22.2). This recommended that the noise measured inside premises should not exceed certain levels for more than 10% of the time.

There should be improved consideration given to noise by Local Authority planning departments when deciding upon planning permission for new buildings. The site location, the topographical features, sound sources, frequency and output levels should always be known and considered. Industrial buildings should be designed, constructed, equipped and maintained in the best possible way to minimize noise emission levels. The work processes, conditions and periods of operation should also be known. It may be necessary to include noise insulation in the building specifications in order to reduce the neighbourhood noise created by new developments. The Buchanan principle for environmental areas, where people take precedence over buildings and traffic, should be generally accepted as part of the planning policy of all LAs.

CHAPTER 23

Future Developments

The study and assessment of noise problems is very complex. Noise can arise from many sources and to a varying extent it impinges upon the lives of every individual for varying periods of their lives. Today there is increasing noise nuisance, and there are indications that more people are reaching the limits of their tolerance to noise.

The number of statutes and regulations that have been brought into use since the late 1960s does indicate that Government is more aware of the problems of noise. However the amount of noise that affects people in their lives is only being reduced very slowly. One reason for this is certainly the system of control and monitoring that operates at present. The responsibility is divided between various agencies to operate controls that are neither efficient nor consistent, because there are only recommended standards. Legislation can only be effective if supported by clearly defined maximum permissible noise levels and standard methods of measurement. It is clear that more research is required to define satisfactory noise levels and methods of measuring them. If control must be a divided responsibility, then adequate training and experience must be provided for the police and Local Authority and industrial environmental officers who carry out important monitoring checks of existing noise levels. There is also a considerable lack of information about the short and long term effects of noise upon people in specific occupations, in their homes, and when enjoying recreation and leisure time.

Noise should no longer be regarded as inescapable and inevitable, whether it is hazardous or disturbance noise. However it must be accepted that noise cannot be eliminated entirely from our modern technological environment. The immediate need is to develop an increasing awareness of noise by the general public, workers, employers, the designers and manufacturers of machines and equipment, and town planners. An interesting project was carried out in Darlington in 1976–78 sponsored by the DOE. This was a local urban noise survey that aimed to make the 97 000 population aware of noise, to assess public reaction, and to examine the problems involved in noise reduction. At present the need for noise abatement often only arises as a result of people's complaints. People do not complain unless they realize that noise is a nuisance and deleterious to the quality of their lives. An increased awareness of noise should lead to pressure upon all concerned to reduce noise to tolerable and acceptable levels. Noise reduction has a relatively low priority in government policies, in industry, and in research and development programmes. What is needed in the future is an agreed policy of noise abatement, and a widespread determination to carry this out by all persons and authorities in all environmental situations.

References

Barr, J., *The Assault on our Senses*, Methuen, 1970.

Brooks, P. T., *Problems of the Environment*, Harrap, 1974.

Bugler, J., *Polluting Britain*, Penguin, 1972.

Chicken, J., *Hazard Control Policy in Britain,* Pergamon Press, 1975.

Consumers' Association, *Aircraft Noise* (*Which—August 1974*), Consumers' Association.

Department of the Environment, *Pollution: Nuisance or Nemesis*, HMSO, 1972.

Department of the Environment, *Protection of the Environment*, HMSO, 1972.

Duerdon, C., *Noise Abatement*, Butterworth, 1970.

Open University, *The Control of the Acoustic Environment, PT272 Units 11 and 12*, O. U. Press, 1975.

Noise Advisory Council, *Neighbourhood Noise*, HMSO, 1971.

Noise Advisory Council, *Noise in Public Places*, HMSO, 1974.

Noise Advisory Council, *Noise in the Next Ten Years*, HMSO, 1974.

Noise Advisory Council, *Noise Units*, HMSO, 1975.

Noise Advisory Council, *Traffic Noise etc.,* HMSO, 1972.

Royal Commission on Environmental Pollution, *First Report*, HMSO, 1971.

Royal Commission on Environmental Pollution, *Fourth Report*, HMSO, 1974.

Taylor, R., *Noise*, 2nd Edn, Penguin, 1975.

SECTION VI

Pollution Control and Legislation

CHAPTER 24

Legislation, Implementation, and Monitoring of Pollution

24.1 UK Policy and Implementation

Any system of pollution control must be based upon a framework of legislation and regulations. These can take the form of Acts of Parliament, or EEC Directives which the UK ratify. At a different level, the UK is party to a number of international agreements, particularly for the control of marine pollution and radioactivity. Once statutory acts have become law they permit specific orders and regulations to be made to cover particular pollution control needs. For example Local Authorities can make Smoke Control Orders for their own districts under the powers of the Clean Air Act 1956, subject to confirmation by the Department of the Environment. The present UK policy for the control of pollution is based upon a framework of broadly based legislation, which has a built-in flexibility to allow variation of implementation.

The amount of pollution, and the type and severity, varies in different localities, for example particular rivers, towns, and industrial plants may show excessive water or atmospheric pollution. Because of this situation, successive governments in the UK have adopted the policy that the primary responsibility for pollution control should be in the hands of Local or Regional Authorities. Central Government produces legislation, but the implementation is delegated to Local Authorities or agencies, except for radioactive radiation. This divided responsibility has both advantages and disadvantages. The policy is in accordance with the British system of decentralized local government, under which there is local control of defined areas of administration, services, and finance. The concept of devolved responsibility is based upon the belief that local and regional authorities know their own parochial problems, and they should be allowed to deal with them within a framework of national legislation. The continuance of this two-tier system assists in the acceptance of control measures. Indeed attempts by various governments to restrict local responsibility have always met with strong criticism and opposition from authorities. The pollution control policy does allow flexibility, and the authorities in any one area may exercise discretion in the limits of pollution that are imposed, to take account of local needs, resources, social priorities, and the capacity of the environment to absorb pollutants. Therefore the chief advantage is that the British system is more acceptable, and it provides a better way to improve local pollution conditions than a national mandatory policy, applied equally across the country irrespective of the varying levels and types of pollution occurring. But one disadvantage of a flexible system is the local variation of interpretation of what constitutes a pollution hazard, and what environmental standards are required. The absence of national statutory standards for pollution control, and a common method of enforcement, will always cause variation in the degree of pollution abatement achieved. The UK policy is at variance with the views of most of the EEC member countries, and has been the subject of considerable criticism. The British philosophy is stated in the DOE Paper No. 11, *Environmental Standards*, 1977 as follows: 'Environmental controls should depend on the balance between the benefit to be gained by the use of a particular substance, the costs of control, and the risks involved in its use or in its introduction to the environment'. This statement introduces the element of cost into pollution control. Undoubtedly the implementation of national control standards would be much more expensive for Central Government, Local Authorities and private industry, than the present system. Whatever system is in use, it should achieve the ultimate aim of any pollution control method, to protect the population in their place of work and in the wider life environment from hazardous substances released into the biosphere.

Monitoring of pollution is fundamental to pollution control. The term monitoring is used in several ways, but generally it means the repeated measurement of pollution concentrations to enable changes to be observed over a period of time. In a more restricted sense, the term is used to describe the regular measurement of pollution levels in relation to some standard, or in order to assess the effectiveness of control measures. Monitoring is carried out to obtain information about current levels of harmful or potentially harmful pollutants in waste discharges to the environment, present in the air, water, and soil of the environment, or in plants, animals and people. The information obtained enables an assessment to be made of the extent of pollution, the rising and falling levels of specific pollutants, and the control measures in operation.

This assessment can be made by comparing monitoring measurements with standards. These are predetermined values of a specific concentration of a pollutant that should not be exceeded. There are four different types of standards in use for various purposes. Probably the most frequently used is the emission standard. This type prescribes the maximum of solid, liquid, or gas that can be discharged or emitted into the environment, and usually applies to emissions into the atmosphere and water. But there are also emission standards for noise and radioactivity. Biological standards prescribe the permitted levels of accumulated pollutants that must not be exceeded in a biological system, for example the concentration of mercury in fish. Production standards set the level of pollutants that must not be exceeded in relation to manufactured products (for example the lead content of petrol, or additives in foodstuffs), or they specify design characteristics of products, such as the noise level for motor vehicles. Lastly, environmental quality standards prescribe the maximum levels of pollution that must not be exceeded in a given area or medium, however many sources of emission are present (for example the noise emission from all the machinery in a workshop). Some standards are expressed in specific terms. Maximum permissible levels (MPLs) are used for water and radioactivity, for instance the maximum dose rate of radiation, or the maximum concentration of a radioisotope. Alternatively, threshold limiting values (TLVs) are the maximum concentration of pollutants to which people at work should be exposed without causing adverse health effects. TLVs and MPLs are usually expressed as parts per million (ppm), or milligrams per litre (mg/l), or grams per cubic metre (g/m^3). A standard is formulated by an appraisal of the known scientific data, clinical effects, and damage to body tissues of a given exposure to a pollutant. When this has been done, the standard value is set at a level which will prevent damage to health, for exposure to existing and likely long term concentrations of the pollutant. It is not easy to set standards because of the varying physiological response of different organisms, and the accumulation of pollutants in body tissues. Damaging effects may be produced in a short or long period of time, according to the internal body effects and the rate of accumulation. Standards can be determined for animals under varying environmental conditions, but there are many more uncertainties in deciding levels for people. No absolute standards can be prescribed, so they must be periodically reassessed and changed in the light of experience. National standards are not prescribed in UK legislation, but they may be included in regulations, or codes of practice, or recommendations, or local and Regional Authority consents for discharges and emissions.

In the UK, surveys and regular monitoring are carried out by a number of Government Departments and agencies. About 70 different bodies and establishments are involved working under five Ministries, and details of the types of monitoring being carried out are given in the appropriate chapters that follow. In addition to all these operations by separate agencies, there are two national surveys carried out. These are the National Air Pollution Survey (NAPS) continuously carried out mainly by LAs and coordinated by the Warren Spring Laboratory, and the periodic River Pollution Survey coordinated by the DOE. Clearly monitoring is complicated, and overlap and duplication of responsibility occurs for some pollution areas. All programmes and the information obtained need to be coordinated and assessed, and this is nominally the responsibility of the DOE. This Department has a Central Unit on Environmental Pollution (CUEP), which is responsible for keeping under review current pollution measures and all monitoring programmes, the appraisal of new problems and needs, and the cooperation of briefing and representation at international pollution meetings. The CUEP is assisted by five sectorial Monitoring Management Groups responsible for air, fresh water, marine waters, biological health, and land pollution. These Management Groups should ensure that all monitoring programmes are scientifically designed to produce useful information on existing pollution levels and long term trends. The five Groups are assisted by two cross-sectional groups concerned with the biological and medical aspects of the monitoring assessment scheme. The five Management Groups are the DOE mechanism to assess the human health and ecological risks associated with specific pollution levels, and the concentrations of existing and new pollutants. The CUEP works in parallel association with Government research establishments, universities, and hospitals, which carry out sponsored pollution research programmes. The DOE and the CUEP also liase and consult with national advisory bodies concerned with the pollution aspects of clean air, water, waste management, nuclear safety, and noise.

Pollution is not confined to any one country or continent, because waste discharges to the atmosphere, rivers, and oceans spread widely over the globe. International cooperation is more and more necessary as the amount of world-wide pollution increases, especially in regard to monitoring, the assessment of pollution effects, and information exchange and research. There are a number of international bodies and programmes concerned with pollution, and the UK contributes and participates in some of this work. During the 1970s a considerable effort has been made towards establishing international cooperation on pollution problems. All these initiatives should be welcomed, but many of the problems being investigated require long term study, and have yet to show their benefits and effectiveness.

24.2 International Cooperation

The pollution control measures and monitoring that are operative in the UK are only part of the British anti-pollution effort. Chemical pollution of the atmosphere and water, waste disposal, oil leakages, the use of pesticides, and radiation pollution are international and world-wide problems. Therefore it is essential that there is international liason, exchange of information, and cooperation aimed at the joint use of similar control standards by many countries. Much cooperation exists at present through international and bilateral agreements, national participation in international surveys, monitoring programmes and research. The UK is involved in the work of the EEC, the OECD, the UN Environmental Programme (UNEP), the UN Economic Commission for Europe (ECE), NATO, and conventions concerned with marine pollution.

The European Economic Community

The EEC Environmental Programme was initiated in 1972, and its broad aims are to improve the quality of life, the surroundings and living conditions of the Community population. The programme operates in two parts. The first part is concerned with the reduction of pollution and nuisances, and includes studies on human health risks, the economics of pollution, and the production of suitable standards for specific pollutants. The second part of the EEC Environmental Programme aims to produce an improvement of pollution problems in urban areas, and in the working and natural environments. The established EEC procedure is to draw up proposals for pollution control measures, which are then considered by the Council of Ministers and the European Parliament. When adopted the proposals become directives, which EEC member countries accept and implement as legislative measures.

There are a number of directives that have been approved, and others that are under consideration for adoption in the future. For air pollution, there are directives and proposals covering particulates, sulphur dioxide, the sulphur content of fuel oils, the reduction of lead in petrol, and quality standards for lead in the environment. The EEC has proposals to deal with land pollution, and these include the areas of general waste disposal; the disposal of toxic, agricultural, oil, quarrying, and mining wastes; and proposals for encouraging recycling within the Community. For pest control the EEC is considering directives for residues in fruit and vegetables, and standardization of the classifying, packaging, and labelling of pesticide preparations. The Community also hopes to produce proposals whereby European pesticide manufacturers can obtain clearance of their products in respect of safety and efficiency, and this directive would operate in parallel with existing national pesticide approval schemes. There are water pollution directives or proposals to improve the quality of water used for drinking or bathing, the levels of biodegradability of detergents and other potentially hazardous pollutants, and the control of specialized industrial wastes such as those discharged from pulp mills. The EEC has proposed directives to implement the Internation Conventions on Marine Dumping, and for pollution from land based sources. The Community is also party to the Bonn Agreement whereby the North Sea coastal nations will cooperate to provide early warnings, and assist each other in dealing with major marine oil spillages. Through EURATOM, the EEC has adopted a directive for basic safety standards of radioactivity that ensure that the ICRP standards are applicable to all EEC members. In regard to noise pollution, the EEC has directives to implement common maximum noise levels for motor vehicles, construction and certain other machinery. These directives should ensure that imported vehicles and machinery within the EEC countries conform to common noise standards.

Another aspect of EEC pollution work is a pilot study for the European Chemical Data and Information Network (ECDIN). This is to design a data bank of environmentally significant chemicals from which information can be rapidly made available to all members for pollution control. ECDIN will be compatible with the UNEP International Register of Potentially Toxic Chemicals (IRPTC). As all the EEC directives become fully implemented throughout the Community there should be a considerable improvement in the quality of the environment, and advances made towards the achievement of international pollution standards.

The Organization for Economic Co-operation and Development

The OECD are a group of major western industrialized countries which discuss common problems and coordination of policies. The Organization set up an

Environmental Committee in 1970 in accordance with its policy of making economic growth qualitative as well as quantitive. The OECD member countries conduct coordinated investigations into such projects as eutrophication, detergents, the control of certain hazardous chemicals, the long range transference of air pollution, the environmental effects of motor vehicles, traffic restraint, and trans-frontier pollution. Although these projects parallel some of the EEC work the emphasis is different, because the OECD proposes solutions to problems that take into account economic and other factors. The OECD also has a subsidiary body called the Nuclear Energy Agency, which is concerned with the marine dumping of nuclear wastes amongst other considerations.

The North Atlantic Treaty Organization

NATO established a Committee on the Challenges of Modern Society (CCMS) in 1969 to carry out international studies of specific problems of the human environment. Each study is carried out under the leadership of one member country, and they have included the pollution of inland and coastal waters, air pollution, advanced water treatment, and the land-fill disposal of hazardous wastes. The UK has taken the lead in the last two studies. The aim of CCMS is to promote the exchange of technical knowledge and expertise between members, which can then be used to improve their environments.

The United Nations

The UN Environmental Programme (UNEP) was set up on a global scale following the UN Conference on the Human Environment held at Stockholm in 1972. The UK is one of the 58 supporting countries, and the DOE is responsible for the British contribution to the programme. In 1976, UNEP designated their priority areas of work as human settlements and health, ecosystems, environment and development, oceans, energy, and natural diasters. UNEP is a very wide ranging programme, unlike the more restricted work of the EEC, OECD and NATO.

The UK is especially involved with two UNEP initiatives. The first is 'Earthwatch', which includes the establishment of an International Referral Service for Sources of Environmental Information (IRS), and the Global Environmental Monitoring System (GEMS). As well as contributing to the work of IRS and GEMS, Britain is establishing the UK National Referral System (NRS), and Data Network on Environmentally Significant Chemicals (DESCNET). The second UNEP initiative, to which the UK contributes, is the International Register of Potentially Toxic Chemicals (IRPTC). This register is to provide data on environmentally hazardous chemicals, regulatory measures, criteria, and standards. IRPTC will facilitate data exchange on a global scale and provide rapid information for early warning and action in relation to specific pollutants. The UK is also interested in the human health and ecosystems programmes under UNEP.

In 1971, the UN Economic Commission for Europe set up a body of Senior Advisers on Environmental Problems to assist the ECE member governments. The members of the Commission include all European countries, the USSR, and North America. The advisers have given priority to projects concerned with the impact assessment of the environment, the integration of environmental policies into socio-economic planning, non-waste technology, air pollution, noise, energy, and transportation. Additional work is being carried out on information systems and data exchange, land uses and planning, solid and toxic wastes, and biodegradable substances.

The UN established the International Atomic Energy Agency (IAEA) in 1957, and this advises 100 member countries. The Agency produces basic radiation standards in line with ICRP, and recommendations for the disposal of radioactive wastes, the transportation and supply of fissile materials, and the necessary protective safeguards. A UN Scientific Commission on the Effects of Atomic Radiation was set up in 1955, and this reports on the radiation levels from various sources and provides scientific evidence of human radiation effects.

The UN World Health Organization is concerned with pollution and safeguards to human health. The Organization has published international and European standards for the maximum permissible levels of pollutants in drinking water, and upper limits for radioactivity. The UN Codex Committee of Pesticide Residues has the responsibility for proposing tolerances for pesticide residues in specific foods, and it is a subsidiary body of the WHO and FAO.

Other International Co-operation

There are numerous international conventions for the control of marine pollution, to which the UK is party. Some of these are described in Chapter 27. Briefly, there are conventions dealing with the intentional discharge of oil and chemicals, the taking of preventive action against coastal oil pollution, the size of oil tankers, the liability for and the payment of compensation for oil clean-up operations, the dumping of potentially harmful wastes at sea, the prevention of marine pollution from land based sources, and the preservation of the marine environment (UN Law of the Sea Convention). The International Commission on Radiological Protection (ICRP) consists of a group of twelve scientists of which one-third are British at present. The ICRP is responsible to the International Congress of Radiology, and it is independent of the UN and other international organisations. Part of the work of the ICRP is to produce recommendations for basic standards for the protection of radiological workers and the general population, as well as secondary standards for particular isotopes. These standards are universally accepted by all countries using radioactive materials, and are therefore one of the few examples of global standards for pollution control.

The World Meteorological Organisation has an air pollution programme to measure air turbidity, and the chemical content of air and rain. Some designated UK meteorological stations participate in the WMO programme, and also supply data on rainfall acidity and the total sulphur content of the air for OECD studies.

This brief summary some of the international pollution programmes that exist, shows that the UK is actively participating in many of them. Britain is also deriving benefit from the collation of data and the formulation of standards. Some of the UK standards in use at present have been derived from the recommendations of the WHO, FAO, IAEA, and ECE.

CHAPTER 25

Control of Atmospheric Pollution

The use of coal was first recorded in England in 852. By 1273 the use of coal in London was prohibited because it was 'prejudicial to health'. The first smoke control order may be said to date from 1306, when a Royal Proclamation prohibited artificers from using sea coal in their furnaces. One offender is on record as having been executed, but smoke pollution from sea coal continued in London to at least the end of the sixteenth century. In 1661, John Evelyn submitted a treatise to Charles II in which the author recognised the hazards of smoke pollution, and proposed in effect the first concept of a smokeless zone, but no action was taken. The development of the Industrial Revolution in the nineteenth century, and the use of steam engines and furnaces, caused Parliamentary Select Committees to study air pollution in 1843 and 1845. As a result, Smoke Abatement Acts covering the London Metropolitan area were passed in 1853 and 1856. Shortly afterwards a Royal Commission was set up to inquire into hydrochloric acid pollution emitted from alkali works using the new Leblanc process. This resulted in the first Alkali etc. Works Regulation Act in 1863, and the setting up of the Alkali Inspectorate to implement the act. In 1875, a Public Health Act was passed containing a smoke abatement section which is still used as a basis for present legislation. The first Local Authority smoke inspectors were appointed in Birmingham in 1903, and in Leeds in 1905. Concern about smoke pollution continued to increase in the late nineteenth and early twentieth centuries, and various abortive attempts were made to introduce legislation in 1887, 1913 and 1914. After the First World War, the Newton Committee studied domestic smoke pollution and reported in 1921. In 1926, the Public Health (Smoke Abatement) Act was passed and this amended the previous Public Health Acts of 1875 and 1891. In 1935, the first completely smokeless housing estates were recorded, and the 1936 Public Health Act included the provisions of the 1926 Act. The first smokeless zone was proposed for Manchester in 1946, and a zone was established in Coventry in 1951. The following year the infamous London smog caused real concern, and the Beaver Committee on Air Pollution was set up in 1953 and reported in 1954. The result of this was that in 1955 the whole of the City of London was declared a smokeless zone, and the Clean Air Act was passed in 1956. Despite this, a further smog incident caused 750 deaths above the normal rate in Greater London in 1962.

Legislation

The Public Health Acts of 1936 and 1969 include provisions for LAs to deal with nuisances causing general air pollution. These are described as 'any dust or effluvia caused by any trade, business, manufacture, or process, and being prejudicial to the health of, or a nuisance to, the inhabitants of the district'. For example, the processing of animal residues is declared an offensive trade under the act. LAs can make nuisance orders under the Public Health (Recurring Nuisances) Act 1969 to prevent the recurrence of specific types of air pollution.

The Clean Air Acts 1956 and 1968 replaced and extended the parts of the Public Health Acts that dealt with smoke nuisances. The two acts include provisions to cover the emission of dark smoke, grit, dust, and fumes emitted from boilers and furnaces in domestic, commercial, and non-registered industrial premises. Control of chimney stack heights is included to facilitate adequate dispersal of emissions. Under the 1956 Act, LAs can make smoke control orders, subject to confirmation by the Secretary of State, whereby it is an offence to permit smoke missions, unless they are caused by the use of authorized fuel or smokeless fuel, in an authorized fireplace (furnace, grate, or stove). Grants can be given to the occupiers of premises to offset the cost of replacing heating appliances. The Clean Air Acts cover domestic premises, colliery spoil banks, railway engines, ships and vessels in inland waters and estuaries, and the burning of waste in the open. The acts are administered by the Environmental Health Officers (EHOs) of the LAs, and cover about 300 000 premises of various types of England and Wales. There are no standard permitted emission limits prescribed under the acts, but the DOE issues guide lines for the use of EHOs. In this respect LAs act autonomously, but there is some liaison between HMACI inspectors and EHOs at local level. The former provide advice and assistance regarding non-registered premises, and the latter pass on complaints regarding registered premises. The national coverage of pollution by smoke

Table 25.1.   Premises covered by smoke control orders in England and Wales, September 1975

Region	Premises (thousands)	Percentage target achieved
Northern	424	43
Yorkshire and Humberside	1023	61
North West	1357	63
West Midlands	626	39
East Midlands	374	37
South East	482	39
South West	95	33
East Anglia	22	12
Greater London	2805	93
Total for England	7208	59
Total for Wales	10	30
Overall Total	7218	59

control orders is very variable. (Table 25.1). It is evident that since the implementation of the 1956 Act, only a 59% coverage has been achieved over the 19 year period. Also the distribution of the smoke control areas is very irregular, varying from 93% in Greater London to under 40% in the industrial Midlands. More progress needs to be made at local level to reduce the commonest type of atmospheric pollution caused by smoke and sulphur dioxide.

Industrial air pollution is partly controlled by the Alkali etc. Works Regulation Act 1906, which applies to certain chemical and industrial processes that may cause pollution hazards. The Alkali Act requires the process, not the works, to be registered, and by 1976 about 2200 works and 3700 industrial processes were registered and subject to inspection by Her Majesty's Alkali and Clean Air Inspectorate (HMACI). The control criterion used by the inspectors under the act is 'to use the best practicable means of preventing the emission into the atmosphere from premises of noxious or offensive substances, and for rendering harmless and inoffensive such substances as may be so emitted'. The term 'best practicable means' is crucial to the implementation of the act, and is interpreted to mean reasonably practicable having regard amongst other things to local conditions and circumstances, to the current state of technical knowledge, and to financial implications. Therefore, except for four processes specified under the act, HMACI do not set standard emission levels for specific similar processes. The inspectors collaborate closely with industry, and carry out a policy of confidential cooperation and joint investigation aimed at improving emission standards.

Pollution from road vehicles is covered by the Motor Vehicles (Construction and Use) Regulations 1973, made under the Road Traffic Act 1972. These regulations embody the principle that exhaust emissions should be partly controlled by the design and manufacture of vehicles. They require new vehicles to be so constructed that no avoidable smoke or visible vapour is emitted. Also no person shall 'use a motor vehicle on the road from which any smoke, grit, sparks, ashes, cinders, or oily substance is emitted if the emission causes or is likely to cause damage to any property, or injury or danger to any person who is actually or reasonably expected to be on the road'. Therefore the regulations require the vehicle user to maintain and operate the vehicle efficiently to reduce exhaust pollution. However it only applies to new vehicles made after 1973, and not those that were produced before this date. The Road Traffic Act 1972 also stipulates an annual road test or 'M. O. T.' for all vehicles of over 1.5 tonnes unladen weight, and this includes smoke emission. Authorized examiners are empowered to carry out spot checks at the road side, and tests on the operator's premises for smoke emission. Since 1974, all diesel engined vehicles must conform to a British Standard on smoke emission. The UK vehicle manufacturers have also been forced to redesign exhaust systems to reduce the emission of carbon monoxide and unburnt hydrocarbons, to conform to the US anti-pollution regulations. EEC directives also require a reduction of unburnt hydrocarbon emission of 35%, and carbon monoxide emission of up to 30%, and these stipulations apply to the UK as a Community member.

The Control of Pollution Act 1974, Part 4, concerns atmospheric pollution and came into operation in 1976. It extends the powers of LAs beyond the Clean Air Acts, and enables them to carry out air pollution investigations and obtain emission data from any premises other than private residences. The information must be available to the public so that they may be aware of particular premises that may be sources of pollution. The 1974 Act also allows regulations to be made to control the content and composition of motor vehicle fuels, and the sulphur content of oil fuels.

Control of radioactive substances discharged into the atmosphere is included in the Radioactive Substances Act 1960, and emissions are authorized by the UKACI on behalf of the DOE and the MAFF. The Health and Safety at Work etc Act 1974 includes the provision to secure the health, safety, and welfare of persons at work, and to use the best practicable means to prevent emission into the atmosphere of noxious or offensive substances. This act also provides for the monitoring of atmospheric conditions and emissions by employers.

Implementation

The control of statutory legislation is the responsibility of four government departments, and implementation is carried out by various agencies (Table 25.2). The summary below shows that the control of atmospheric pollution is a very fragmented and divided responsibility. The control of noxious fumes, grit, and dust discharged into the atmosphere from registered premises is carried out by HMACI. This provides the most consistent and standardized control, although there is some variation in standards to take account of local conditions. Atmospheric pollution from non-registered industries, and domestic and

Table 25.2. Atmospheric pollution control by government departments in E & W

Department	Responsibility
Air Pollution	
DOE	Pollution from registered premises
	Advice to LAs on all other types of premises
DTp	Motor vehicle pollution
DT	Aircraft pollution
DI	National survey of air pollution, and specialized surveys
Radioactivity	
DEn	Policy, treatment and storage of wastes
	Safety of nuclear power plants and inspection
DOE	Policy on waste management, and disposal on land
	Disposal of waste effluent
	Discharges to the atmosphere
	Pollution from the use of radioactive substances
MAFF	Discharges into water, and effects on the aquatic environment

commercial premises in controlled by LA Environmental Health Officers. This local type of control produces considerable variation in standards, despite advice provided from Central Government and liason with HMACI. There are no UK national standards for air quality, but proposals towards attempting to achieve this objective are under consideration by the EEC, and if acceptable would apply to Britain.

The emission of exhaust fumes from some road vehicles is controlled by DTp examiners who can carry out roadside checks of heavy goods vehicles. Such vehicles must also undergo annual tests at heavy vehicle testing stations, and the operator's licence is not renewed if the exhaust emission is unsatisfactory. However, neither of these controls applies to cars or light vans under 1.5 tonnes unladen weight, and this is an important weakness in the control of vehicle pollution. The emission of lead into the atmosphere as a result of using petrol with lead additives is being controlled by DEn regulations for petroleum fuels. There is no control exercised over the smoke trails from aircraft that usually occur on take-off, or the exhaust fumes from railway diesel traction units. Both these types of exhaust fumes are not considered to provide any significant contribution to lower atmospheric pollution. The control of radioactive emissions into the atmosphere is the responsibility of HMACI.

To assist in the formulation and administration of legislation, government departments receive advice from statutory bodies, and they promote research. Under the Clean Air Act 1956, the Government established the Clean Air Council which was responsible for keeping under review the progress made in abating air pollution. Many of the recommendations produced by the Clean Air Council were incorporated in Part 4 of the Control of Pollution Act 1974. It is regrettable that the DOE announced in September 1979, that the Clean Air Council was to be abolished in order to reduce public expenditure. Research is carried out at the DI Warren Spring Laboratory (WSL) and this establishment is particularly concerned with the dispersion, long range transport and reactions of air pollutants, offensive odour nuisances, and specialist local surveys. The DOE Transport and Road Research Laboratory (TRRL) at Crowthorne, Berkshire carries out work on vehicle pollution. Other specialized work is carried out by NERC, the ARC and the MRC. The general trend of research sponsored by the Government is to investigate the impact of air pollution, its extent, and the behaviour of pollutants after emission. Work on abatement and control techniques is mainly carried out by trade associations, or individual companies in accordance with the general principle that the polluter pays.

The Radiological Protection Act 1970 authorized the establishment of the National Radiological Protection Board (NRPB), which is responsible to the DHSS. The Board advises other government departments, carries out research, and produces codes of practice in consultation with the UKAEA. Radiological research may be grouped into two main areas. Biomedical work is carried out by the MRC and the NRPB. The former is concerned with basic research, and the latter undertakes applied investigations for the protection of industrial workers.

Biological work is the province of the MAFF, and mainly consists of research into the environmental pathways of radioactive substances in biological organisms and food stuffs.

Monitoring

The measurement of air pollution at source is desirable because once pollutants are emitted into the atmosphere their movement, concentration, chemical reactions, and effects are largely controlled by the prevailing environmental conditions. The monitoring of emissions at source, and a knowledge of the weather conditions allows a rough calculation of the atmospheric concentrations to be made. From this it is possible to develop quality objectives and emission standards for specific pollutants, to provide a measure of protection for the population at risk. Regular monitoring is necessary to provide a check upon the compliance with regulations, and to assess whether changed atmospheric conditions require a revision of the emission standards. The scale of monitoring that is carried out varies very widely, and may consist of atmospheric sampling over the whole country, or a region or town, around a factory or power station, at the roadside, or at a very limited number of sites. The most extensive monitoring carried out in the UK is the National Air Pollution Survey (NAPS) which has been in operation on a comprehensive scale since 1960. The survey is based upon the measurement of smoke and SO_2 concentrations over a 24 hour period at 1300 sites. Approximately 1100 sites are located in selected urban areas, and there are 200 county based sites to sample the background concentration of smoke and SO_2. Some 600 sites also measure grit and dust emissions. NAPS is carried out by 500 LAs, selected meteorological stations, and CEGB power stations. All the sampling data is sent to the Warren Spring Laboratory where it is checked, processed, and published.

Table 25.3. Monitoring of atmospheric pollution in E & W

Department	Responsibility	Agency
DOE	Smoke and SO_2	LAs, CEGB, WSL, Met Office
	Grit and dust	LAs, CEGB, WSL
	Industrial emissions	Industry, HMACI
	Lead, fluoride, cadium, etc	Industry, LAs, HMACI
	Metals generally, acidity	LAs
	Motor vehicle emissions	WSL, TRRL
DI	Nitrogen oxides, oxidants, hydrocarbons	WSL
	Airports	WSL
	Organopesticides in rain and the atmosphere	Laboratory of the Government Chemist
MAFF & DES	Trace elements in rain and soil	NERC, AERE, MAFF
MOD	Chemicals in air and rain	Meteorological Office

Other monitoring programmes are less extensive and usually only involve a small number of sites. They are concerned with a range of different pollutants and particularly hazardous substances such as metals, pesticides, oxidants, etc., as shown in Table 25.3. The overall coordination of these surveys is usually the responsibility of HMACI or the WSL.

Monitoring of Radioactivity

Radioactive pollution of the environment can originate from two main sources. Radioactive dust may be discharged into the atmosphere as a result of nuclear weapon testing, accidents at nuclear reactors, or from nuclear fuel processing plants. Radioactive low-level waste is discharged into rivers and the sea, and it is assumed that the subsequent dilution will prevent any serious pollution. To safeguard the population it is necessary to carry out monitoring of both types of discharge at source, and measurement of the concentrations of radioactivity in air, water, soil, and the tissues of living organisms.

The control of radioactivity is provided by a system of monitoring that is distinct and separate from the other types used for air, water, and land (see Table 25.4).

The level of fallout radiation is carefully monitored because the whole population is exposed to this, irrespective of any industrial radioactive pollution.

Table 25.4. Monitoring of environmental radioactivity in E & W

Government Department	Responsibility	Agency
DI	Fallout in air and rain	AERE
ARC & DES	Fallout in milk and farm produce	ARC Letcombe Laboratory
DT	Fallout in drinking water	Laboratory of the Government Chemist
MAFF	Fallout in fish and coastal waters	MAFF Fisheries Laboratory
MRC & DES	Cosmic and gamma radiation in air and soil	Leeds University
	Radioactive wastes—all types	Nuclear site operators
DOE	.. —gaseous	UKACI
DI	.. —liquid	Laboratory of the Government Chemist
DEn	.. —stored on site	UKAEA, DEn
MAFF	.. —in air and sea	MAFF Radiobiological Laboratory
DOE	.. —around nuclear sites and from non-reactor sources	Radiochemical Inspectorate

Monitoring is carried out by the DI, DT, MAFF, and DES supported research establishments. Radioactive discharges from nuclear establishments are controlled by authorized limits determined by the DOE and the MAFF. The authorizations are issued after careful consideration of the best practicable means of discharge reduction, the environmental pathways involved, and the population at risk. The operators of licensed sites such as research stations (UKAEA), or nuclear power stations (CEGB) monitor their own discharges and the surrounding environment. Independent checks are carried out at the sites by the DOE and the MAFF. The BNFL fuel processing site at Windscale is subject to more intensive monitoring because of the large quantities of nuclear waste that it produces. Air and ground radioactivity is monitored, and in addition isotope concentrations in water, food, and the built environment are measured at localized sites. Four government departments are concerned with the monitoring of all types of radioactive wastes and their disposal on land, into rivers, and the sea. This diverse control requires close liaison and co-operation between the departments and their agencies, to ensure that a national management system operates efficiently to safeguard the population. There is no sectorial management group for radioactive pollution comparable to the other five groups for other types of pollution. The official reason for this is that it is not considered expedient to separate the different aspects of radioactive pollution.

The possibility of accidental radioactive discharges from nuclear sites has led to the specification of emergency reference levels. These ERLs specify minimum radiation doses, which if reached during an accident, would require counter measures and emergency action. Every nuclear establishment has an emergency reference plan to provide for the complete evacuation of the population living within approximately 1 km of the site during a 2 hour period. Extensive monitoring of the environment, including all foodstuffs growing in the area, would be undertaken before people were allowed to return to their homes. Nuclear fuels and waste are transported over varying distances, and leakages or accidents can occur during transit. These incidents are covered by the National Arrangements for Incidents Involving Radioactivity (NAIR), which are administered by the NRPB. This scheme provides for the rapid summoning of assistance from a national list of organizations and industrial firms, whose personnel have the expertise and resources to deal with the emergency and carry out monitoring.

Radioactive pollution is the only type of pollution in the UK for which there are national standards. Basic primary standards are prescribed in accordance with the recommendations of the ICRP and EURATOM. From these primary standards, secondary standards or derived working limits (DWLs) can be determined, and these state the total quantity of a specific radioactive substance that is permissible in the human body without the maximum allowable dose being exceeded. Tertiary standards are also used for air, water, and food, that specify the concentration of a radioactive substance which would produce a maximum dose in particular circumstances. These standards are expressed as the maximum limit of radioactivity that must not be exceeded, irrespective of the cost of the control measures. Secondary and tertiary standards are based upon evidence derived from tracing the environmental pathway of a specific substance through biological cycles and food nets to the human body tissues.

Prior to 1978, the ICRP recommendation for primary standards of radiation dosage were stated for a working individual, and for the population at large. An individual member of the public should not be exposed to more than 0.5 rem per year. The population of a country should not receive an average total dosage exceeding 1.0 rem per person over their reproductive period of 30 years. In 1978, the ICRP amended its recommended limits to 'the lifetime whole body dose to individual members of the public from all sources of radiation except natural background should not normally exceed 7 rem'. The ICRP no longer recommends average yearly dose limits for the population as a whole, but any control system for radiological protection should ensure that the average annual dose received by the whole population is less than 50 m rem. These new standards are endorsed by the UK NRPB, and a new UK standard has been produced as follows: 'the contribution to this annual average dose (50 mrem) from all UK waste management practices should not exceed 10% of this value or 5 mrem'.

CHAPTER 26

Control of Land Pollution

Legislation

The first legislation concerned with waste disposal was the Public Health Act of 1875, which gave Local Authorities powers in respect of refuse collection and disposal. Since this date, numerous statutes have been made to control specific aspects of waste disposal and pollution, but up to 1972 the legislation was concerned with domestic and trade waste and did not cover industrial waste disposal on land.

The Public Health Act 1936 gives discretionary powers to LAs in England and Wales to collect and dispose of household waste free of charge, and trade waste upon payment. They can also provide plant for treating waste and sell any reclaimed materials. Under this act LAs also have powers to provide litter receptacles in streets and public places, to sweep and water the streets, and empty cesspools and privies. However, the act does not deal with industrial, farm, mining and quarrying waste. The Civic Amenities Act 1967 is concerned with the problem of the disposal of bulky wastes and disused vehicles. LAs are obliged to provide sites for the disposal of this type of rubbish without charge to the public. Some LAs make special collections of bulky wastes, but many rely upon the public transporting the waste to the disposal site. This act makes it an offence to abandon vehicles and rubbish upon open land or a highway. LAs are obliged to remove dumped vehicles, and have powers to remove other waste deposited in the open air, if the land owner does not object. The Litter Act 1958 made the deposition of litter in public places an offence, liable to a fine of £10. The Dangerous Litter Act 1971 increased the maximum fine to £100, and required the courts when convicting offenders to take notice of any risk incurred to persons or animals.

Local Government in England and Wales was reorganized in 1974 as a consequence of the Local Government Act 1972. In England, the new County Authorities are responsible for waste disposal and the operation of refuse plants and tips. But in the Metropolitan Counties, the new District Councils are responsible for the collection of domestic and trade waste and the disposal of bulky wastes. This split responsibility between the two types of LAs in England has been one of the many criticisms of the reorganisation. No doubt the overall planning functions of the County Authorities were seen to include waste disposal planning. In Wales and Scotland, the new authorities are responsible for both the

244

collection and disposal of all wastes, and so they have not changed their functions as a result of reorganization.

The disposal of industrial waste, other than by LAs, was not covered by legislation until 1972. Prior to this date there was much public concern about the uncontrolled and indiscriminate dumping of toxic wastes in various parts of the country. The second report of the Royal Commission on Environmental Pollution 1972 recommended Government action, and so in the same year the Deposit of Poisonous Waste Act was passed as an interim measure. This act makes it an offence to deposit on land any poisonous, noxious, or polluting waste in such circumstances as to endanger persons, or animals, or pollute water supplies. The act attempts to control toxic waste disposal by requiring the disposer to give three clear working days notice, and details of the composition, quantity, and destination of the waste to the waste disposal authority (LA) and the regional Water Authority. Land fill operators who dispose of the waste are also required to confirm receipt and disposal of the notified waste within three clear working days. This mechanism is not a consent procedure, but it does allow disposal authorities to safeguard the environment. The act has been effective in that much more highly toxic waste is now dealt with at specially approved sites such as the one at Pitsea in Essex.

The Control of Pollution Act 1974 must be considered as a very important statute in relation to the control of environmental pollution. The first part of the act, concerned with waste on land, includes many of the recommendations of the Department of the Environment (DOE) Working Parties on the Disposal of Toxic Wastes and Refuse Disposal, which produced reports in 1970 and 1971. The Control of Pollution Act redefines the powers of LAs in respect of household and trade wastes, and includes powers for the collection of industrial waste upon payment, where LAs are required to do this. The private sector industrial waste service can still operate independently. The act establishes a new mechanism for control of waste disposal operations through a site licensing system. All operators, whether in the public or private sectors, wishing to use treatment plants and disposal sites must apply for a licence from the County Local Authority acting as the Waste Disposal Authority. The licence consent will include site operating conditions, and the Authority has extensive supervisory and enforcement powers over the operations carried out on the site. The Waste Disposal Authority must also ensure that satisfactory facilities are available for waste disposal. They must carry out a survey of the present sites in use, the types and quantities of waste to be disposed of, and produce a plan showing the methods of disposal and the control of operations that are to be used. The plan must include all types of waste including domestic, trade and industrial types, it must be published for comment and sent to the DOE. Special or hazardous wastes will be subject to stringent regulations, and the DOE will issue technical memoranda and advice to Authorities about the handling and disposal. When the act is fully implemented the Deposit of Poisonous Waste Act 1972 will be repealed.

Five other statutes are also concerned with land pollution. The Town and Country Planning Act 1962 enables County Planning Authorities to control change of land use, and the activities of the extractive industry. All excavations and open cast mining operations are subject to planning permission which can include conditions regarding spoil tipping and land reclamation. Unfortunately there is considerable variation in the stringency with which LAs operate this section of the act in different parts of the country. The Health and Safety at Work etc Act 1974 requires that waste must be disposed of without risk to employees or the public. The Radioactive Substances Act 1960 is concerned mainly with the safe disposal of low level nuclear waste, and not with the treatment and storage of high level nuclides. The act requires the registering of premises, except Crown establishments, which use radioactive substances. This is to ensure control of radioactive waste at source, and to try to prevent a waste hazard arising. The granting of registration can include conditions regarding the labelling and removal of radioactive materials from the premises. An authorization is required to cover the accumulation and the disposal of waste at all registered premises, and this stipulates the type of waste, means of disposal, and any measurement of radioactivity of the waste, or of the environment, that may be necessary. Registration and authorizations are granted by the DOE and MAFF in England, but LAs are consulted and informed of the details. The premises within scope of the Radioactive Substances Act are educational and research establishments, the UK Atomic Energy Authority (UKAEA), and licensed nuclear sites such as nuclear power stations. High level nuclear wastes are controlled by separate legislation. The Atomic Energy Authority Acts of 1954 and 1971 include provision to control the wastes produced by the UKAEA and British Nuclear Fuels Ltd (BNFL) through authorizations given by the DOE and MAFF. These two Ministries have to consult LAs, Water Authorities, local fishery committees, and other public or local bodies when making authorizations for waste disposal. The Nuclear Installations (Licensing and Insurance) Act 1959 applies the same controls to nuclear power stations.

Legislation also covers the safe use of pesticides and other similar chemicals by agricultural and horticultural workers and the general public. The Health and Safety at Work etc Act 1974 obliges employers and employees to take reasonable care of their own health and safety at work, and not put others at risk as a result of their activities. This act also empowers Ministers to make regulations to prohibit or regulate the manufacture, supply, keeping, or use of any substance. More specifically, the Health and Safety (Agriculture) (Poisonous Substances) Regulations 1975, which are made under the 1974 Act, are designed to protect operators using toxic pesticides from poisoning; 60 active ingredients are named, and 26 operations are described, where the operators must wear specified protective clothing. The labelling of pesticides intended for sale is covered by the Farm and Garden Chemicals Act 1967. The Farm and Garden Chemicals Regulation 1971, made under the 1967 Act, require farm and garden pesticides to be labelled clearly with the names of the active ingredients they contain, and to bear prescribed warnings about their use in relation to humans and animals.

In addition to legislative requirements, any risks in the use of pesticides are, hopefully, minimized by the implementation of non-statutory arrangements.

The MAFF, the Department of Health and Social Security (DHSS), and the Department of Employment Health and Safety Executive (HSE), jointly operate the Pesticides Safety Precautions Scheme or PSPS. These government departments are advised by the Advisory Committee on Pesticides and other Toxic Chemicals, and they exercise a close supervision over the introduction of new chemical substances. Under the voluntary PSPS scheme, manufacturers submit information to the Ministries in respect of the properties, persistence, toxicity, proposed use, and method of application of the new product. If satisfactory, then PSPS recommendations are issued for the safe use of the pesticide, and the protection of users, consumers of treated crops, and wild life. Under the scheme, the use of any pesticide can be reviewed if it is subsequently shown that a new risk has arisen. There is also the Agricultural Chemicals Approval Scheme (ACAS), that covers all crop protection chemicals which have been cleared under the PSPS. The scheme is operated by the Agricultural Chemicals Approval Organization on behalf of the MAFF, and approves chemicals for their biological efficiency. The 'A' symbol of approval enables advisers to recommend, and users to select, appropriate crop protection chemicals for use against specific pests, and the scheme is intended to discourage the use of unsatisfactory products.

Implementation

The above summary of land pollution legislation shows the diverse pattern that has been developed. Perhaps this is because land pollution is less coherent compared to atmospheric and water pollution. The present control system involves three government departments, various agencies under government or private control, and voluntary schemes operated by manufacturers. Broadly the DOE is responsible for land planning, waste disposal, and land pollution generally; the MAFF is responsible for controlling all types of pollution in food, agricultural chemicals, and farm wastes; and the Department of Energy controls radioactive waste processing and storage on nuclear sites (Table 26.1).

Table 26.1. Land Pollution Control* by Government Departments in E & W

Department	Responsibilty
DOE	Disposal of toxic wastes
	Advice on refuse collection and disposal
	Disposal of industrial waste
	Mining and quarrying waste
MAFF	Pesticide control
	Disposal of farm wastes
	Crop protection chemicals (with ACAO)
MAFF, DHSS DE (HSE)	Pesticide approval (with PSPS)

*For pollution by radioactivity see Table 25.2.

Control of the disposal of hazardous wastes is delegated to Local Authorities with strengthened resposibilities under the Control of Pollution Act 1974. The implementation of the licensing system for hazardous waste disposal will help to reduce the potential pollution of water courses, and safeguard water supplies. One anomaly of the 1974 Pollution Act is the failure to include agricultural wastes produced from intensive livestock units and silage silos within the category of industrial waste. Pre-discharge treatment of this highly organic waste is not usually carried out on farms, so that potential pollutants such as pesticides, herbicides, and nitrates are discharged into the environment. If farm waste disposal was subject to the same licensing system as other industrial waste, then agricultural land and water pollution could be more controlled. Also under the 1974 Act, LAs for the first time must survey the disposal of all types of wastes, and produce plans for the siting and control of waste disposal operations. This requirement is very necessary, and it is regrettable that the present Government announced in September 1979 that the implementation of this section of the act had been deferred for an indefinite period.

Specific legislation for pesticides only covers agricultural and horticultural workers, and the labelling of proprietary products. Although the MAFF can prohibit or regulate the manufacture of certain hazardous products, this is rarely done. The British way is for the Ministry to persuade manufacturers to conform to voluntary restrictions, as occurred in respect of DDT and some organochlorine compounds. The only other control of pesticides is through the non-statutory voluntary PSPS scheme, which attempts to limit the chemical content of proprietary preparations and provide advice upon their use. There is no control of the quantity of pesticides that are sold, or how they are used by farmers, horticulturalists, and the general public. Many individual cases of misuse occur each year, and consequently the concentration of some pesticides in the environment is reaching very high levels.

The storage and disposal of radioactive wastes is very closely controlled by the highly specialized national inspectorate of the Health and Safety Executive (HSE). They work closely with the licensed nuclear site operators such as BNFL at Windscale, the CEGB, and the UKAEA.

There is no specific legislation to control land dereliction, other than the requirement placed upon the National Coal Board and mineral extraction companies to restore the land despoiled by open-cast mining and quarrying. The redemption of derelict land, spoil heaps, waste tips, and inner urban dereliction may be carried out by LAs, but this is a non-obligatory operation. Local Authorities are encouraged to do this through Central Government grants, but apparently many LAs remain oblivious to the quality of the environment of their citizens. There is no mention of land dereliction in the all-embracing Control of Pollution Act 1974, which is indicative of the Government attitude to this aspect of land pollution.

Government departments seek and receive advice about land pollution from three advisory committees at present, which consist of a representative spread of interests and technical experts. The DOE and DI jointly set up a Waste

Management Advisory Committee in 1974. Its terms of reference include waste management policies, the best use of resources, the safe and efficient disposal of waste, the interrelationship between waste utilization and disposal, the recycling of specific wastes, and research and development. A second committee is jointly responsible to the MAFF and the DES. This is the Advisory Committee on Pesticides and other Toxic Chemicals, which is supported by a scientific sub-committee of experts and a panel for the Collection of Residue Data. The Panel is responsible for organizing the monitoring of pesticide residues in foodstuffs, which is carried out by six laboratories. The DEn has a Nuclear Safety Advisory Committee that provides independent advice upon nuclear safety matters.

The central administration and control of pollution must be supported by research, which is vital to identify potentially hazardous pollutants, and for establishing methods of control. Research being carried out in relation to land pollution can be grouped into four main areas of work, namely waste management, pesticides and herbicides, foodstuffs, and radioactivity. Research on waste treatment, recovery and recycling is carried out at the DI Warren Spring Laboratory, and the AERE at Harwell offers a hazardous materials service of technical information and advice to LAs and industry. Pesticide research is undertaken in five UK laboratories, including the Monks Wood Experimental Station, and the ACR Experimental Station at Rothamsted, whilst pesticides and herbicides are investigated at the ACR Weed Research Association. Analysis of foodstuffs for pesticide residues is coordinated by a sub-committee of the Advisory Committee on Pesticides and other Toxic Chemicals. The work is carried out in six laboratories: DI Laboratory of the Government Chemist; MAFF Plant Pathology Laboratory, Harpenden; MAFF Pest Infestation Laboratory, Slough; MAFF Fisheries Laboratory, Lowestoft; DAFS Scientific Services Laboratory, Edinburgh; and the DAFS Freshwater Laboratory at Pitlochry.

Monitoring

Pollution is not directly transferred from the land to people, except in the case of dusts and direct contact with toxic materials. Pollutants deposited on land usually enter the human body through the medium of contaminated crops, animals, food products, or water. Land pollution can also damage terrestrial ecosystems, resulting in a deterioration of the conservation and amenity value of the environment. There are four main types of land pollution monitoring proposed in the UK. Local Planning Authorities are required under the Control of Pollution Act 1974 to record all areas of land that are known to affected by pollution, and to supply the information to the DOE. Land areas include solid waste tips of all types, including toxic wastes, and the location of spoil heaps. LAs are also responsible for surveying and recording the amount of derelict land in their areas, and making proposals for reclamation schemes. The third type of monitoring involves measurement of the levels of hazardous substances on polluted land. This includes levels in soil, general background levels, and the

Table 26.2. Monitoring of land pollution* in E & W

Government Department	Responsibility	Agency
DOE	National survey of waste disposal facilities, and pollution risks	LAs, WAs
DOE	Survey of notifications of deposits of poisonous wastes	UKAEA Hazardous Wastes Service
MAFF	Amounts and effects of pesticides, heavy metals and slurry disposal	MAFF control laboratories
DI	Pollution in soil, flora, fauna, birds' eggs, etc	Laboratory of the Government Chemist
ARC and DES	Determination of pest residues and dietary contamination	Agricultural and horticultural research stations
NERC and DES	Pollution in birds and eggs, lead pollution on roadsides, soil analysis, mercury in ecosystems	Nature Conservancy, Monks Wood Station, Universities

*For pollution by radioactivity see Table 25.4.

assessment of conditions around known sources of industrial pollution such as manufacturing plants, solid waste and sludge tips. These measurements are carried out by LA environmental officers, the MAFF and other agencies. Complementary to this type of monitoring, there is the measurement of pollutant levels in plant tissues, terrestrial animals, human foodstuffs, and man. This is also carried out by the MAFF, as well as by the DI and other agencies. The present monitoring system can be summarized as shown in Table 26.2. This system of monitoring is again decentralized, and so it is not surprising that some overlap exists between the various monitoring programmes. For example the amount of pollutant and its effects in plants, animals, and food stuffs are investigated and assessed by the ARC, NERC, the Nature Conservancy, and the MAFF laboratories. Cooperation between these bodies is operated by the five CUEP Sectorial Management Groups which include the Land Pollution Monitoring Group. This group is particularly concerned with the deliberate or accidental dumping of potentially hazardous wastes, the effect of pollution on land use planning, and the use of waste on agricultural land. As previously mentioned, there is a real need for periodic national surveys of the extent and types of all derelict land. The former Ministry of Housing and Local Government, and later the DOE, has initiated annual surveys of derelict land by LAs. However these have been restricted surveys that mainly only included certain specified types of land which justified reclamation, and for which grants may be available. These surveys were also not comparable, because the DOE changed the requirements and survey methods in 1972.

The monitoring of biological organisms, soil, and water for pollutant concentrations can be carried out, and the results assessed against prescribed standards. But there are some types of land pollution where no quantitative standards can be applied, for example dereliction, degradation of the natural rural environment, and reduction of amenity and recreational facilities. Consequently there are inevitably many variations in the views and *ad hoc* standards used by LAs and other bodies.

The system of voluntary control of pesticides and herbicides in use in Britain is stated by Government to be satisfactory. But this system only provides a partial control of the use of these hazardous chemicals. There is no control of the quantities of the approved products that are used, and it is this situation that is causing national and international concern. For example, it is becoming common practice for farmers to use a battery of several pesticides rather than individual substances to control so-called resistant pest species. The Royal Commission on Environmental Pollution 7th Report 1979 comments upon the voluntary PSPS and ACAS schemes. Whilst reluctantly agreeing to the continuance of a voluntary scheme, the Commission proposes some changes to improve the control and use of pesticides, fungicides, and herbicides. The PSPS and ACAS schemes should be amalgamated, and consideration given to the single scheme becoming statutory. Some change will probably be required soon, because the EEC are considering proposals for an acceptance directive to cover a scheme for the safety and effectiveness of pesticide chemicals throughout the Community. The Royal Commission Report proposes consideration of a licensing scheme for pesticide operators, whether they are farm workers or aerial spraying contractors. Before a licence is granted the operator would have to provide evidence of training and proficiency in crop spraying. This proposal is aimed at preventing excessive chemical treatment of crops through incorrect techniques or unnecessary spraying. To further control this increasing malprac-tice, the report also proposes an increased responsibility for the MAFF Agricultural Development Advisory Service (ADAS). This advisory body should provide more guidance to farmers and horticulturalists, tackle pollution problems, and check current husbandry practices as sources of pollution. The Commission emphatically seeks a change of attitude and approach to the present use of agricultural chemicals of all types. It states that the MAFF should change its somewhat defensive and protective attitude to farmers, and actively initiate action to reduce the use of agricultural and horticultural chemicals to a minimum, consistent with agricultural objectives. In the future, the official policy should be to recognize that agricultural advances in husbandry and techniques must be tempered more to environmental considerations, and the effects that they produce.

CHAPTER 27

Control of Water Pollution

Legislation

There are numerous statutes concerned with water pollution and these have been produced as and when there was a need for legislative action. Broadly the statutes can be grouped into four categories which deal with sewage collection and treatment, discharges to inland waters and the sea, and radioactive substances.

Inland Waters

Probably the first water pollution legislation was the Gas Works Clauses Act 1847 which prohibited the discharge of gas wastes into streams. There was increasing concern about water pollution during the latter half of the nineteenth century, and a Royal Commission on River Pollution was set up in 1865. Eleven years later, a River Pollution Prevention Act was passed in 1876, and this was the starting point of sewage treatment in the UK. During the next century more than a dozen acts were passed dealing with water pollution in rivers and the sea. The Public Health Act 1936 requires sewage disposal authorities to make 'such provision, by means of sewage works or otherwise as may be necessary, to effectively deal with the contents of their sewers'. Under this act it is an offence to discharge into a public sewer any 'prohibitive matter'. This is described as anything that may damage the sewer, or interfere with the flow or treatment process, and includes liquids with a temperature of over 43°C and petroleum spirit. The disposal authorities must carry out their functions without causing a nuisance to the public. The Public Health (Drainage of Trade Premises) Act 1937 is concerned with industrial discharges, and permits them to be released into a public sewer subject to the consent of a Water Authority. The consent or licence includes quality and quantity conditions to safeguard the sewerage system and treatment works, and serves as a means of controlling pollution. The Water Authorities are empowered to levy a charge for the conveyance and treatment of the discharge, and this is based upon its volume and content. The Public Health Act 1961 amended the 1936 Act, and is notable for including farm wastes and drainage under the heading of trade waste for the first time. These three acts provide the statutory basis used for the control and treatment of wastes entering the sewage system. They have been slightly amended by the Control of Pollution Act 1974.

The Rivers (Prevention of Pollution) Act 1951, the Clean Rivers (Estuaries and Tidal Waters) Act 1960, and most of the Rivers (Prevention of Pollution) Act 1961 have been repealed under the Control of Pollution Act 1974. The Salmon and Freshwater Fisheries Act 1975 prohibits the putting into water of any matter that is poisonous or injurious to fish or their spawning grounds. It also permits Water Authorities to make by-laws to regulate waste discharges into water that are detrimental to any freshwater fish. This 1975 Act is the latest of a number of statutes made since 1861 to protect freshwater fish from the effects of pollution.

The Water Resources Act 1963 is concerned with underground waters. It contains sections to control pollution entering any well, pipe, or borehole through discharges. The act also provides for emergency powers to deal with accidental pollution of rivers, and allows water undertakings to acquire land, where this is necessary to prevent the pollution of reservoirs and underground strata.

Prior to 1974 in England and Wales, sewage disposal and treatment was carried out by about 1400 Local Authorities, public water supplies were provided by 200 water undertakings, and inland waters were managed by 29 River Authorities. The Water Act 1973 reorganized this entire network of control under 1629 different undertakings, and set up nine regional Water Authorities (WAs) in England and one for Wales, effective from April 1974. The Water Authorities are responsible for all aspects of the water cycle. They are responsible for public water supplies and treatment, sewage collection and treatment, control of river and aquifer pollution, land drainage and flood prevention, and the improvement and development of fisheries. The WAs carry out water conservation, and have powers to develop the best use of water space for navigational, recreational, and amenity purpose as advised by the Water Space Amenity Commission. The Water Act also established a National Water Council to advise on national water policy, and to assist and coordinate the efficient functioning of the regional Water Authorities. For example in 1978, it was announced that the National Water Council was to set new standards for water quality, and review all present consents for industrial and farm discharges into streams. The reorganization of the water industry in 1974 coincided with the coming into force of the Control of Pollution Act, which provides the statutory basis for the WAs to exercise pollution control.

The Control of Pollution Act 1974, Part 2 is concerned with water pollution, and most of this section should be fully implemented by the end of 1979. The act covers virtually all forms of water pollution including control of discharges into inland waters. For the first time there is control of discharges into tidal waters and the sea, and discharges from mines. The act makes it an offence to discharge 'any poisonous, noxious or polluting matter . . . . any matter which impedes the proper flow of water of a stream and any solid matter . . . . which is likely to lead to a substantial aggravation of pollution due to other causes, or the consequences of such pollution'. The term stream is defined as 'any river, watercourse or inland water which is natural or artificial or above or below ground'. The Act states that all discharges of trade and sewage effluent into water are subject to consent or disposal licences granted by a Water Authority. Before a consent is granted, the discharger must supply details of the location, nature and composition, temperature, daily flow, and maximum rate of the discharge. The WAs themselves must obtain a consent for their own sewage discharges from the Department of the Environment. The WAs are required to keep a register of all consents, and this must be open for inspection by industry and the general public. The register must record details of the conditions imposed, and of sampling analyses carried out on the effluent and the receiving water, for each consent. Under the Control of Pollution Act, no raw sewage may be discharged from any type of boat into a river or estuary. The WAs also have a duty to take any necessary action to restore the flora and fauna of a stream. The act provides for the imposition of severe penalties on offenders found guilty of causing pollution, and they can be charged with the cost of pollution abatement. The full implementation of this act should eventually result in a significant reduction of water pollution in the UK.

It is interesting to note that some individuals have a right to take action against river pollution under common law. The owner of land fronting on to a river has riparian rights that entitle the owner to a water flow down to the land 'without sensible alteration in quality and character'. Similarly the owners of underground water, fisheries, oyster beds, etc. can take action against polluters under common law. Courts can grant an injunction restraining the continuance of pollution, and they can award damages. Some successful prosecutions against firms and sewage works have been brought under the riparian rights common law.

Marine Waters

The United Nations definition of marine pollution is 'the introduction by man, directly or indirectly, of substances or energy into the marine environment resulting in such deleterious effects as harm to living resources, hazard to human health, hindrance to marine activities including fishing, impoverishment of quality for the use of sea water, and the reduction of amenities'. The UN view is that it is reasonable to make use of the natural capacity of the marine environment to accept and degrade potential pollutants. Therefore control of marine pollution should take account of this and legislation should be made accordingly.

The control of sewage and industrial discharges into coastal or controlled waters in the UK is within scope of the Control of Pollution Act 1974. Controlled waters are defined as 'the sea within 3 nautical miles (5.56 km) from any point on the coast. . . . such other parts of the territorial sea adjacent to the UK as are prescribed. . . .'. However, the control of other potentially polluting matter, such as oil and chemical substances dumped at sea outside territorial waters, comes under other legislation. These British statutes are mainly associated with various international agreements or conventions.

Several acts are concerned with the pollution aspects of petroleum, and oil and gas exploration on the UK continental shelf. The Petroleum (Production) Act 1934 allows licences to be granted for the exploration and exploitation of oil and gas, and these require companies to prevent the escape of waste oil into the sea. The Continental Shelf Act 1964, and the Petroleum and Submarine Pipelines Act 1975 are both concerned with oil pipelines. Consents are required to lay pipelines, and these must be properly constructed, correctly routed, and buried where practicable to protect them from corrosion and avoid the risk of damage and oil spillage.

The dumping of various wastes at sea was hopefully controlled under a voluntary code of practice until 1974. The Dumping at Sea Act 1974 requires dischargers to obtain dumping licences from the Ministry of Agriculture, Fisheries and Food (MAFF) for certain potentially polluting substances, as specified by the Oslo and London International Conventions of 1972.

Other aspects of marine pollution are mainly concerned with shipping, and are controlled by a number of international agreements which have been made since 1954. The UK, as well as other maritime nations, has ratified these agreements, and in some cases they have been included in national legislation. The Prevention of Pollution of the Sea by Oil Convention, covering the North Sea and most of the eastern Atlantic, was agreed in 1954, and amended in 1962 and 1969. The 1969 agreement prohibits the discharge of oil in polluting quantities anywhere at sea within 80.5 km of land. The Prevention of Pollution from Ships Convention 1973 extended the 1969 agreement to include the prohibition of discharges of non-persistent oils, dangerous chemical residues, sewage, and garbage. The Oil Pollution Damage Convention 1969, and the Intervention on the High Seas in Cases of Oil Pollution Casualties Convention 1969, are concerned with accidents at sea that cause oil pollution. Both these conventions came into force in 1975, and they give coastal nations the right to take action to avoid or minimize oil pollution threats to their coastline. Two international agreements control dumping at sea and these were signed in 1972. The Oslo Convention on Dumping covers the north-east Atlantic area including the seas around the UK, and the London Convention has global coverage. Both these conventions were ratified by the UK in 1975, and they require dumping at sea to be authorized by national authorities. They place restrictions on the wastes that can be dumped, and prohibit the deposition of radioactive wastes, chemical and biological warfare agents, oils, cadmium, mercury, and organo-halogens. Some chemical substances such as arsenic, lead, copper, cyanides, and fluorides can only be dumped in specific quantities in specially approved locations, where it is assumed that they will be rapidly diluted and dispersed. When all these conventions have been ratified by a large number of countries, there will be a sound framework available for the control of some aspects of marine pollution. The main problem of the future is the systematic implementation of these control measures. This requires the adoption of responsible attitudes by shipping companies and individual ship's captains, and constant monitoring of pollution control.

Radioactive Substances

The discharge of low level radioactive wastes is controlled by the Radioactive Substances Act 1960. Under the act, all users of isotopes, except Crown establishments, must be registered with the Department of the Environment. The DOE and the MAAF exercise national control through a system of authorizations to dispose of wastes. The authorizations specify the type of waste, the means of disposal, any conditions to be observed, and any measurements of radiation levels that may be necessary. Crown establishments are controlled through their administrative regulations. Some low level wastes in sealed, concrete-lined drums are dumped on the sea bed. This is a carefully controlled exercise usually carried out once per year under the control of the UK Atomic Energy Authority (UKAEA) and the MAAF, by international agreement. The operation takes place under the auspices of the Nuclear Energy Agency, which is a subsidiary of the Organisation for Economic Cooperation and Development (OECD). The waste is dumped in a mutually selected area of sea that is clear of the continental shelf, and all the dumped quantities are fully recorded. On land, a national disposal service is available to radioactive waste producers, and this is operated by the UK Atomic Energy Research Establishment (AERE), Harwell, and British Nuclear Fuels Ltd (BNFL). High level wastes are produced by the nuclear power industry and these are stored on site. The processing and storage of high level wastes is controlled under the Nuclear Installations Acts of 1965 and 1969. The Health and Safety at Work etc. Act 1974 provides for general protection of people at work from radiation hazards, and for the prevention of health risks for the general public.

Implementation

Water pollution control is administered by four Government departments and various agencies. Briefly, the DOE has overall responsibility for national policy and coordination, the MAFF is responsible for freshwater fish and marine pollution, the Department of Trade for ship pollution, and the Department of Energy for pollution from nuclear power and oil installations (see Table 27.1).

The government departments receive advice and assistance from a number of central advisory bodies in relation to water pollution. The Water Act 1973, which created the regional WAs, also established the National Water Council. Amongst its functions in regard to fresh water, the Council is responsible for advising on schemes to prevent waste, misuse, or the contamination of water. There are Standing Technical Advisory Committees, responsible to the DOE and the National Water Council, for assistance and advice in respect of water quality, the disposal of sewage sludge, synthetic detergents, biodegradability, water reclamation, and water analysis. There is also a Joint Committee on Medical Aspects of Water Quality which has been set up by the DOE and the DHSS. In the area of marine pollution, there is a Standing Committee on Pollution Clearance at Sea that advises the DOE on the clearing-up of oil pollution.

Table 27.1. Water pollution control* by government departments in
E & W

Department	Responsibility
Freshwater	
DOE	Water management policy
	Sewage effluent disposal, protection of water supplies, and water quality
	Technical advice to LAs and WAs
	National River Pollution Survey
MAFF	Control of pesticides
	Advice on the disposal of farm wastes
	Effects of pollution on fish
Marine	
MAFF	Policy, and fishery safeguards
	Dumping of wastes at sea
DOE	Disposal of solid, and toxic wastes at sea
	Chemical and oil pollution on beaches, and clean-up operations
DT	Oil pollution at sea
DEn	Pollution from off-shore oil and gas operations

*For pollution by radioactivity see Table 25.2.

A considerable amount of research on water pollution is carried out by Government agencies. The MAFF laboratories research into pollutive effects on freshwater fish, the toxicity of oil dispersants and oil, and the behaviour and effects of pollutants reaching the sea. The DI Warren Spring Laboratory is currently investigating new methods of clearing oil at sea and on the shore, and it advises coastal LAs about this problem. The DOE Central Water Planning Unit commissions research by specialist bodies into topics such as water quality, and ecological studies of rivers and estuaries. Additional work is carried out by non-governmental bodies and establishments. The water industry set up a Water Research Centre in 1974 as an industrial research association. The centre carries out research for the DOE, and contract work for members, including the ten regional Water Authorities in the UK. Specialist pollution research on such topics as the effects of heavy metals, and organochlorine compounds is carried out by the Institute for Marine Environmental Research and by NERC. The Nature Conservancy Council (NCC) also carries out research on the pollution effects upon wild life.

Monitoring

The extensive monitoring of water pollution is a divided responsibility carried out by the regional WAs, individual industries, and Government field and research stations and laboratories (see Table 27.2). Freshwater monitoring is carried out at the main points along the flow path of a pollutant. The WAs take samples and make measurements at the pollution source, of the effluent

Table 27.2. Monitoring of water pollution* in E & W

Government Department	Responsibility	Agency
Freshwater		
DOE	River and lake pollutants	WAs
	Trace metals, organic compounds in river water	Water Research Association
DT	Effluents	Laboratory of the Government Chemist
MAFF	Treatment of farm waste before disposal	MAFF experimental stations
	Eutrophication	
NERC & DES	Specialist local surveys, and monitoring	Universities and research stations
Marine		
MAFF	Pollution residues in fish, and shellfish	MAFF Fishery Laboratories
DT	Pesticides in fish	Laboratory of the Government Chemist
NERC & DES	Pollution in estuaries	Universities and research stations
	Specialist local surveys	

*For pollution by radioactivity see Table 25.4.

discharge to sewers and rivers, and where river water is abstracted or enters the sea. Additional monitoring is carried out for pollutants present in river water and the river bed, and the tissues of aquatic organisms. Industrial effluents are monitored by the industries concerned, and checked by the WAs to ensure that they conform to the conditions of the discharge consent. The WAs are also responsible for water quality, and so they regularly sample and measure the BOD, temperature, pH, suspended solids, nitrogen content and turbidity of waters. Special monitoring for heavy metals and potential carcinogens is carried out in addition. The WAs have a duty to make periodic surveys of the flora and fauna and the natural aquatic balance of rivers and streams. Public water supplies must be regularly analysed for their bacteriological and chemical content, both before and after treatment, and this applies to rivers, reservoirs, and underground waters. When the Control of Pollution Act 1974 is fully implemented, the WAs will have a new requirement to monitor waste discharges into coastal waters. The DOE and the DES commission special monitoring programmes for pollution at selected sites, and these are carried out by the Water Research Association, NERC, and the UKAEA, supported by the MAFF research stations, and the Laboratory of the Government Chemist. Particular attention is given to the concentrations of heavy metals, organochlorine pesticides, nitrates, and radioactive isotopes that accumulate in the tissues of fish and other food animals.

Marine pollution monitoring is carried out mainly by the MAFF Fisheries Laboratories at Lowestoft, and they assess the various pollutants in sea water, mud, silt, fish, and shellfish. More specialized work on pesticide residues and PCBs in sea birds and fish is carried out by the DT Laboratory of the Government Chemist. The monitoring of radioactive pollution is also quite extensive. Discharges from nuclear research and fuel treatment establishments, and nuclear power stations are monitored initially by the various site operators, but independent checks are carried out by the MAFF and the DOE to safeguard the environment. Coastal waters receive low level radioactive discharges, and so there is regular monitoring of inshore fish and other marine foods as well as the shore areas. The determination of radioactivity levels in marine foodstuffs is carried out by the MAFF Fisheries Radiobiological Laboratory.

The most comprehensive water monitoring carried out in the UK is the River Pollution Survey, which officially started in 1958. These periodic surveys are made by the WAs and LAs and they involve the sampling of 40 000 km of rivers and canals, and then classifying these water courses according to their water quality. Details of the results of these surveys are given in Chapter 14. A survey of foul sewage discharges into coastal waters was made in 1972 and this was updated in 1976.

CHAPTER 28

Control of Noise Pollution

Legislation

Legislation to deal with noise has existed for a number of years by virtue of noise being considered to be a nuisance. This legal term is not clearly defined by statute but in relation to noise there are three recognized types of nuisance, i.e. public nuisance that inconveniences or endangers the public, private nuisance that interferes with another person's enjoyment of his land, and statutory nuisance as designated by a statute. Prior to 1960, the only action that could be taken against noise was to show that damage or interference had been caused to the beneficial use of premises or personal comfort. Certainly noise was not recognized as a form of pollution.

Since 1960, there have been a number of statutes promulgated to cover road traffic, aircraft, neighbourhood, and occupational noise. The first specific legislation, under which LAs could take action, was the Noise Abatement Act 1960. This was concerned with neighbourhood noise, including that from factories, sports stadia, places of entertainment, construction and demolition sites; and general community noise in residential areas and public places. The act gave LAs powers to make by-laws, and so the control of noise was very variable over the whole country. For example, by-laws exist for the control of transistor radios, hawkers, street noise, acoustic bird scarers, model aircraft, and discotheques. These by-laws are inconsistent in the maximum permissible noise levels that they prescribe, and the Home Office only provides guidance to LAs. Under the Noise Abatement Act, loudspeakers can only be used for specific purposes between 08.00 and 21.00 hours, and ice cream vans can only use loudspeakers when on the move during 1200 to 1900 hours. Noise in licensed premises can be controlled through the granting of licences by local magistrates, and they also control the issue of entertainment licences for dance halls, cinemas, and discotheques. Outdoor motor sports frequently cause complaints about noise nuisance. Spectator sports such as motor racing and hill climbing are subject to local planning permission, and come within the scope of the Noise Abatement Act 1960. This act proved to be very restrictive and complicated to implement. Action could only be taken after the noise nuisance had occurred, and there were no effective powers available to produce a lasting reduction in environmental noise levels. The Noise Abatement Act was repealed by Part 3 of the Control of Pollution Act 1974.

Table 28.1. Noise limits for manufactured road vehicles

Vehicle Type	UK noise limits for vehicles in use after 1.11.70 (dBA)	UK noise limits for vehicles in use after 1.11.74 (dBA)	EEC directive 1970 (dBA)	EEC directive 1974 (dBA)
Motor cycles over 125 cc	86	—	—	—
Petrol-engined cars	84	80	83	82
Diesel-engined cars	84	82		
Light passenger and goods vehicles	84	82	84	
Heavy vehicles up to 200 b.h.p.	89	86	90	89
Heavy vehicles over 200 b.h.p.	89	89	92	91
Land tractors	89			
Buses with capacity of over 12 passengers	89			

Road traffic noise has caused increasing concern and complaint since 1905, but no effective legislative action was taken until after the Wilson Committee Report 1963. The Motor Vehicle (Construction and Use) Regulations were first made in 1973, under the Road Traffic Acts. These regulations require an efficient silencer to be fitted, maintained in efficient working order and not altered, and prohibit the use of vehicle horns between the hours of 23.30 and 07.00. They also prescribed maximum permissible noise levels for all types of vehicles then in use, and for new vehicles licensed and in use after 1.11.1970. Further regulations were made in 1974 as shown in Table 28.1. The table shows the UK regulations and the EEC directives of 1970 and 1974. Overall the EEC limits are higher than those of the UK, but the latter have a more comprehensive coverage and require lower limits for heavy lorries. It should be noted that the above noise limits do not apply to the millions of vehicles manufacturered and in use, prior to 1970. Home Office statistics show that in 1970, there were 13 327 successful prosecutions for vehicles with faulty silencers, but the total of prosecutions for all other noise offences was only 760. The total annual prosecutions of 14 087 must be considered in the context of about 14.5 M vehicles licensed in 1970. It is interesting to note that the Ministry of Transport (MOT) road test for vehicles was in operation for 16 years before it required an examination of the vehicle's exhaust system, dating from 1976. It is obvious that the enforcement of vehicle noise regulations has been far from rigorous. Vehicle testing and the measurement of noise emitted whilst in use on the road was prescribed in 1969, but its implementation is said to cause practical difficulties. The police are responsible for spot testing vehicles on the road, but few tests are actually done, due to police reluctance and cost. The Road Traffic Regulation Act 1967 allows Local Authorities to make traffic regulation orders prohibiting or restricting the use of roads to some or all traffic. This act has been reinforced by the Heavy Commercial Vehicles (Controls and Regulations) Act 1973, whereby the same type of order can be applied to heavy vehicles. These two acts can be used to reduce road traffic noise and vibration in specific urban and residential areas. The Road Traffic Act 1974 includes a compulsory national type approval scheme. Commencing in 1977, manufacturers have to submit new types of vehicles to the DT for testing in respect of safety and environmental standards, including exhaust systems. All new vehicles sold must conform to the approved type, and random checks can be made including exhaust noise output.

The Land Compensation Act 1973, and the Noise Insulation Regulations 1973 made under the act are measures to help in the reduction of noise in buildings. The regulations allow grants to be made by highway authorities, towards the cost of building sound insulation against traffic noise from new and improved roads. The Land Compensation Act also allows LAs to acquire land for the erection of earth embankments and acoustic barriers to reduce traffic noise.

Participation sports, including rallies and trials on public roads, are covered under the Motor Vehicles (Competition and Trials) Regulations 1969. Under the regulations, control is operated through the Royal Automobile Club (RAC), which authorize the event and prescribe a noise limit of 3 dBA below the statutory requirements for all competing vehicles. The General Development Order 1973 operates for other events on privately owned sites, for example grass track racing, auto-races, minor hill climbs, and go-karting. The Autocycle Union is the official controlling body, and they specify a maximum level of 100 dBA at a distance of 3 m.

Aircraft noise control was excluded from the Noise Abatement Act 1960, but this type of noise is covered by other legislation. The Civil Aviation Acts of 1949 and 1968 require all aircraft landing and taking-off in the UK to have a noise certificate containing specific requirements. The Civil Aviation Act 1971 allows the Secretary of State for Trade to designate airports for noise abatement measures. To date, Heathrow, Gatwick, Stansted and Prestwick airports have been so designated. The noise measures prescribe minimum noise routes, rates of climb, flying heights, and restrictions on the night flights of jet-engined aircraft. The owners of non-designated airports, such as Luton and Ringway, are responsible for noise control, and they make their own regulations. The measures for both types of airport are agreed with the Civil Aviation Authority (CAA). The Air Navigation (Noise Certification) Order 1970 requires new types of certificate of airworthiness to be made for all aircraft coming into service from 1.1.1976 onwards. These certificates must ensure that new aircraft are half as noisy as the pre-1969 types of similar weight. This is a welcome attempt to reduce the nuisance of aircraft noise at airports. The grant-aiding of sound insulation costs for residents in the vicinity of designated airports is allowed by the Airports

Authority (Consolidation) Act 1975. Non-designated airports may use Local Authority powers for the same purpose.

There is no specific legislation relating to occupational noise levels in industrial premises, but the Health and Safety at Work etc. Act 1974 requires employers to ensure the health, safety, and welfare of their employees, as far as is reasonably practicable. Under this act, a code of practice—'Reducing the Exposure of Employed Persons to Noise', 1972 operates, and this recommends a general maximum noise level of 90 dBA for an 8 hour day and a 5 day week inside premises. The compliance of employers with the code is controlled by the HSE Factory Inspectorate.

Prior to 1974, the only way to limit outside noise emitted from industrial premises was to prove negligence under common law. This position was changed by the Control of Pollution Act 1974, Part 3, which came into operation on 1.1.76. This act superseded the Noise Abatement Act 1960, but still specifies noise as a statutory nuisance. Local Authority powers have been strengthened and extended in scope, and they have a duty to inspect their areas in order to detect and take action against noise nuisance. Notices can be served on firms or individuals to effect abatement, restriction, or prohibition of noise nuisance. The act also permits the DOE to issue codes of practice to provide guidance to authorities and operators on types of noise which may constitute a nuisance. The Control of Pollution Act 1974 also provides powers to LAs to establish noise abatement zones and prescribe maximum permissible noise levels, where there is excessive noise from factories and other premises. When designating the noise abatement zones, LAs must take account of local conditions, circumstances, and the cost involved, and monitor the noise emission. The act includes a new provision relating to construction sites. LAs can impose requirements regarding plant and machinery, the hours of working, and permissible noise levels. The act does not lay down any national standards for environmental noise, but adheres to the British philosophy that control should be dealt with locally according to the prevailing circumstances.

Implementation

The responsibility for the control and legislation of noise is divided between four Government departments, and implementation and control is devolved to various agencies (see Table 28.2). The complexities of noise require expert research and advice. The Government Departments are advised by various research establishments and bodies including the Transport and Road Research Laboratory (TRRL), the Building Research Establishment (BRE), the National Physical Laboratory, the Royal Aircraft Establishment, the National Gas Turbine Establishment, and the MRC. The chief areas of work are concerned with the quietening of aircraft, subjective response to aircraft noise and its effects upon health and hearing, operational flying techniques to reduce noise, and the effectiveness of noise abatement measures. The DOE is advised by

Table 28.2. Control of noise in England and Wales

Government Department	Administrative responsibility	Monitoring
DOE and DTp	Road traffic and vehicles	Police, DTp inspectors
	Neighbourhood noise	LA EHOs
DT	Aircraft noise	CAA
	Aircraft noise abatement	DT, and airport management
DE	Occupational noise	HM FI, employers

the Noise Advisory Council which was established in 1970. Its terms of reference include 'to keep under review, progress made generally in preventing and abating the generation of noise'. The Council has several working groups of technical experts, and has published reports on aircraft noise, neighbourhood noise, traffic noise, noise in public places, and noise units.

Monitoring

Noise differs from other forms of pollution in relation to standards. There are considerable difficulties in determining the relationships between noise levels, and their effects upon people below the threshold of hearing damage. The effects depend upon the subjective reaction of individual people to the amplitude and the frequency of the noise, the time of day, place, and the background against which the noise is heard. Consequently few standards exist for noise, but there are maximum levels for vehicles, aircraft, and certain types of occupational noise. The Government issues circulars for the guidance of LAs, and these have included advice upon industrial noise, the implementation of the 1974 Pollution Act, and planning and noise. For example, when LAs are planning new residential development, they should site it so that road traffic noise measured outside buildings should not exceed 70 dBA on the L_{10} (18 hour) index. The building specification for residential property should ensure that the L_{10} noise inside buildings is not greater than 40 to 50 dBA. For existing urban areas the circulars recommend that the ambient level should not exceed a corrected noise level of 75 dBA in the day or 65 dBA at night.

Monitoring of noise levels is carried out by various bodies. For example LA Environmental Health Officers are responsible for the measurement of neighbourhood noise, and the HSE Inspectorate monitor occupational noise to check compliance with codes of practice. Employers are recommended to monitor their own premises, and the British Airports Authority maintains a monitoring system for aircraft noise around Heathrow Airport. Vehicle noise monitoring is the responsibility of the Police and the Department of Transport Inspectors.

CHAPTER 29

Future Developments

Chapters 25 to 28 inclusive have described the present legislation for controlling the various forms of environmental pollution. The position has improved markedly over the last decade, and from 1969, twenty-five Acts of Parliament, Regulations and Orders have been legislated by successive UK Governments. One of the most important statutes is the Control of Pollution Act 1974. For the first time, legislators have shown official recognition of the integrated nature of pollution by the inclusion in one comprehensive statute of sections covering waste on land, the pollution of water, noise, and pollution of the atmosphere. Perhaps the main defect of the 1974 Act is that land pollution is only included in respect of the collection and disposal of waste on land and recycling. The other aspects of land pollution, mainly concerned with dereliction and conservation, are not recognised and included. It is regrettable that the UK national economic conditions since 1975 have so far prevented the full implementation of the Control of Pollution Act. This is a situation that should be remedied as soon as possible, because the full implementation of this Act would help to improve the increasing pollution of this technological age.

The implementation of pollution measures is peculiarly British, and its divided and fragmented operation is clearly apparent in the various tables of control and monitoring responsibilities shown in earlier chapters. The Royal Commission on Environmental Pollution 5th Report criticized the divided control and proposed certain changes. The Commission favoured the continuance of shared control by Central Government and the LAs, but expressed doubt about the existence of sufficient technical knowledge and expertise at local level to provide effective pollution control. The Report proposed the establishment of a new central pollution inspectorate with wider responsibilities. The new body, or HM Pollution Inspectorate (HMPI) should encompass the present responsibilities of HMACI and HMFI, but have extended environmental control. HMPI should be responsible for the control of air pollution, noise from industrial processes, and hazardous waste control. It should also collaborate closely, and advise the Water Authorities regarding industrial effluent discharges, and Local Authorities in respect of non-industrial air and noise pollution and nonhazardous waste disposal. HMPI would have no responsibilty for aircraft or vehicle noise. These proposed changes would not alter the LAs responsibility for the control of pollution from non-industrial sources. The Royal Commission further proposed that HMPI should be responsible for controlling all radioactive discharges to the environment in consultation with the MAFF. In addition to improvements in the national control of pollution by inspectorates, there is a need to change the general approach to control as carried out by the various bodies concerned. The Royal Commission on Environmental Pollution 5th Report criticized the tendency of some control agencies, such as the UKACI, WAs, and LAs to operate within a strictly limited area. For example, a London power station is required to wash flue gases to remove sulphur dioxide and reduce air pollution. But if the wash effluent is discharged into the River Thames, this causes deoxygenation and increases river pollution. The two agencies concerned are HMACI and the Water Authority, and these bodies must cooperate to jointly solve this problem of pollution transference. The basis of control methods stated as the 'best practicable means' should be expanded to become the 'best practicable environmental option'. The broad aim of pollution control should be to reduce the total pollution effect upon the environment, and not reduce one type of pollution at the expense of another.

National surveys are carried out for the assessment of atmospheric pollution, and inland water pollution. The Royal Commission on Environmental Pollution 6th Report 1976 proposed that an annual radiation survey of discharges of radioactivity into the air and water, and on to land should be published by HMPI. In addition the NRPB should publish periodic reports of the total radiation exposure of the population. The Report further recommended that the NRPB should have a statutory responsibility to advise the Government upon basic standards and ERLs for radioactivity. If these Commission proposals were implemented there would be improved control of radiation pollution. Also more information about radioactivity would be available to authorities and the public, and this could help to allay, or confirm fears that there is an ever increasing radiation risk to the population.

There are many pollution monitoring programmes being carried out, and again these are operated by many different bodies and organizations. It is necessary to identify the harmful substances being discharged into the environment, and to know the quantities, and the sources from which they originate. But it is also essential to be able to correlate similar monitoring results, so that a coherent picture of the national trends can be realized. Therefore coordination between monitoring programmers is essential, and also cooperation to ensure consistent sampling methods and measurements to produce comparable collected data. The required close coordination is not operating satisfactorily at present. The DOE, CUEP proposals in Pollution Paper No 1 could improve this situation. A designated 'lead laboratory' for each sector of pollution would be responsible for providing guidance on monitoring in regard to equipment and methods of measurement, as well as coordinating the various monitoring programmes within Central Government policy and strategy. For example, the Warren Spring Laboratory could lead on air pollution, the MAFF Fisheries Laboratory on marine pollution, and so on. In addition to this closer organization and control of monitoring, there must also be effective exchanges between the lead laboratories and the establishments carrying out research into the physical and biological systems of the environment.

All the proposals described have been made because of apparent weaknesses in the system of divided responsibility and effort used for pollution control. Close cooperation and overall national supervision of all pollution control work is essential. The responsibility for this at present is centred upon the DOE Central Unit on Environmental Control and its Sectorial Monitoring Management Groups, which were set up during 1975 to 1976. They will need time to fully develop their organization and activities, but it appears that there is need for a considerable improvement in the British system and methods of pollution control in the future.

References

Control of Pollution Act, HMSO, 1974.

Department of the Environment, *Controlling Pollution, CUEP Paper No 4*, HMSO, 1975.

Department of the Environment, *Environmental Standards, CUEP Paper No 11*, HMSO, 1977.

Department of the Environment, *Monitoring of the Environment in the UK, CUEP Paper No 1*, HMSO, 1976.

Department of the Environment, *Pollution Control in Great Britain: How it Works, CUEP Paper No 9*, HMSO, 1976.

Royal Commission on Environmental Pollution, *Fifth Report*, 1976.

Royal Commission on Environmental Pollution, *Sixth Report*, 1976.

SECTION VII

Summary

CHAPTER 30

Pollution in Perspective

The detailed problems of environmental pollution have been discussed in the various sections of this book. But a study of pollution cannot be made in isolation and it must be set within the problems of the global environment in which we exist. Pollution is only one of a matrix of interrelated factors which include world population increase, food, energy and natural resource supply and consumption, developing technology, and economics. It is indisputable that the developing countries all operate a highly developed technological society, with economic pressures for high output and high consumption of goods and services. These countries produce most of the present global pollution, and as this gets worse it is decreasingly being assimilated into the natural environment. This statement summarizes the fundamental pollution problem, substantiated by the historical evidence of developments over the last two centuries.

The Present Situation

The present position in the UK is that some pollution problems are being examined and remedied. Progress has been made over the last 20 years, but it has largely consisted of isolated actions taken in response to the needs and pressures that have arisen, for example smoke abatement, control of toxic waste dumping on land; and the altered use of detergents, asbestos, organochlorine pesticides, and food additives. Other changes such as a reduction of some river pollution and aircraft noise, and some restoration of derelict land areas have taken place. A considerable number of legislative statutes have been enacted, but their effectiveness has been reduced by delayed implementation and lack of consistent and stringent enforcement.

The present attitude to pollution has been succinctly expressed as 'how much pollution can we afford?', implying that the cost of pollution control measures is the controlling factor in pollution policies. Certainly present control policies are usually formulated within the context of national economic policy. The Government view was stated in the CUEP Pollution Paper No 11, 1977 as 'Environmental control should depend upon the balance between the benefits gained by the use of a particular substance, and the costs and the risks involved in its use and introduction into the environment'. From this it follows that estimates of the costs, benefits, and risks involved are required for each control measure.

This is very difficult to carry out because there is a lack of information, and subjective and ethical judgements have to be made in respect of the social and amenity benefits. In fact, some economists have questioned the feasibility and value of cost benefit analysis for pollution decisions. At present many pollution action decisions are taken on the flexible basis of the best practicable means of control in order to achieve a balance of costs and benefits. This procedure is linked to the national policy on control financing, which is often expressed as 'the polluter pays'. This means that the person, or organization creating the pollution should be financially responsible for controlling it, and no specific grants for pollution control are available from Central Government funds. Therefore control measures are not standardized, and can be affected by the ability of the polluter to pay for the cost of implementation.

Perhaps the main achievement of the 1970s has been an increasing, but still limited, recognition of pollution effects and the future environmental implications. There has been more realization of the increasing deterioration of the environment and the quality of life. There has been a growing desire to reduce pollution and improve the environment for future generations. This concern is to be welcomed and encouraged, but it is only a change of attitude shown by some sections of industry and the general public. The present situation is not satisfactory, because there is no significant reduction in pollution taking place within the overall biosphere.

The Future

There are many ideas about how to solve environmental problems. A minority of people propose a reduction of population growth, developing technology, and production and consumer demand; accompanied by the abandonment of economic growth policies. This may be acceptable as a long term objective, but the realization of this requires acceptance on a global scale. At present, no single nation would accept unilateral action along these lines. Conversely, it is advocated that technology and economic growth must continue. In fact this is essential for national prosperity, and to provide the technical expertise and finance to deal with future pollution problems and control as they become apparent. If this latter view is taken as the most realistic and acceptable one, then it can be used as a basis for considering future pollution policies.

Short term pollution abatement measures should continue, because it is essential that acute problems are resolved to safeguard the population from immediate and potential risks. Some short term technical improvements have been discussed in earlier chapters. For example, air pollution could be reduced by improving the efficiency of fuel combustion, by restricted use of the internal combustion engine, and the development of electric or gas power units, or high speed flywheel propulsion. Solid waste could be converted into processed fuel, and used as an energy source instead of being dumped on land. Improved pre-discharge treatment of liquid wastes, and more industrial recovery and recycling, could reduce the water pollution of rivers and coastal waters. Pollution control implementation could be improved if there was better cooperation between the various agencies involved, in respect of the standards used and the stringency of operational control.

Parallel to any short term measures there is a need for long term policies and research studies. One of the most important future pollution hazards is probably the effect upon the population of being exposed to low concentrations of pollutants, and low level radiation, over long periods of time. These pollutive factors enter the body through direct exposure, or they are absorbed through pollution pathways in natural systems and food chains. The long term effects of body accumulations of these substances are often unknown at present, but it is essential to determine the incidence and severity of their effects in the future. The existing lack of knowledge must be rectified by research into the biochemistry of human tissues and cells. It is necessary to know the induced effects of specific chemical pollutants and radiations, so that suitable TLVs can be determined. This long term medical research must be accompanied by systematic monitoring of pollutants, their environmental paths in food chains, and their concentrations in ecosystems. When the target objectives for pollution control are better known in terms of TLVs for specific pollutants, then the necessary implementation measures can be determined, and their effectiveness regularly monitored.

A long term pollution strategy is required, based upon aims that are clear and widely recognized. The human race lives within the earth's biosphere, and the population depend upon it for their continued existence. Man may be able to exercise a degree of control over his reproductive rate, life style, and national productivity, but he cannot completely control the environment. Man can only attempt to modify it to suit his aspirations and objectives. Man's activities tend to create pollution with the possibility of disrupting the natural cycles of the environment. This can directly affect the balanced state of the biosphere. Therefore the disposal of increasing quantities of waste into the environment could seriously upset the stability of the biosphere and that of the earth in the future. Clearly the long term ultimate aim for pollution must be to reduce its effects to a level where there is no potential threat to the biosphere. This means that the quantity and types of pollutants that are necessarily discharged into the environment should be capable of being assimilated, without causing any hazard to human and wild life.

Some people regard the possibility of achieving this long term aim with considerable scepticism, so long as the present economic and technological growth policies continue to operate. They rightly pose the question: 'how can pollution be substantially reduced within the context of increasing diversity and output of waste?'. However, it may be that the sociological effects of advanced technology and the diminution of natural resources in the next century will necessitate a fundamental change in the present growth policies of the developed countries.

The eventual development of pollution and its effects are uncertain in the long term future. But in the next two decades, changes are required to deal with the present pollution situation. Many Governments need to adopt a more positive approach in terms of 'how polluted should the environment be?', instead of the present view of 'how polluted can the nation afford to be?'. There should be general acceptance of the long term aim of pollution reduction, and an unqualified commitment to carry out coordinated policies and provide the necessary resources. All the developed countries will need to educate and motivate their industries and populations to adopt responsible roles in the control and reduction of all forms of pollution.

List of Abbreviations

ACAS	Agricultural Chemicals Approval Scheme
AERE	Atomic Energy Research Establishment (Harwell)
ARC	Agricultural Research Council
BNFL	British Nuclear Fuels Limited (Windscale)
CAA	Civil Aviation Authority
CEGB	Central Electricity Generating Board
CUEP	Central Unit on Pollution Control (DOE)
DE	Department of Employment
DEn	Department of Energy
DES	Department of Education and Science
DESCNET	Data Network on Environmentally Significant Chemicals (UK)
DHSS	Department of Health and Social Security
DI	Department of Industry
DOE	Department of the Environment
DT	Department of Trade
DTp	Department of Transport
E & W	England and Wales
ECDIN	European Chemical Data Information Network (EEC)
ECE	European Commission for Europe (UN)
EEC	European Economic Community
FAO	Food and Agriculture Organization (UN)
GB	Great Britain
GEMS	Global Environmental Monitoring System (UNEP)
HMACI	Her Majesty's Alkali and Clean Air Inspectorate (DE)
HMFI	Her Majesty's Factory Inspectorate (DE)
HSE	Health and Safety Executive (DE)
IAEA	International Atomic Energy Agency (UN)
ICAO	International Civil Aviation Organization

ICRP	International Commission for Radiological Protection
IRPTC	International Register of Potentially Toxic Chemicals (UNEP)
IRS	International Referral Service for Sources of Environmental Information (UNEP)
LA	Local Authority
MAFF	Ministry of Agriculture, Food and Fisheries
MOD	Ministry of Defence
MRC	Medical Research Council
NATO	North Atlantic Treaty Organization
NERC	National Environmental Research Council
NNI	Nuclear Installations Inspectorate
NRPB	National Radiological Protection Board
NRS	National Referral System (UK)
OECD	Organization for Economic Cooperation and Development
PSPS	Pesticide Safety Precautions Scheme
RCI	Radiochemical Inspectorate
UK	United Kingdom
UKAEA	United Kingdom Atomic Energy Authority
UN	United Nations
UNEP	United Nations Environmental Programme
WA	Water Authority
WHO	World Health Organization (UN)
WSL	Warren Spring Laboratory (DI)

Select Bibliography

General

Allen, R., Goldsmith, E., *et al*, *A Blueprint for Survival*, Penguin, 1972
Beckerman, W., *In Defence of Economic Growth*, Jonathan Cape, 1974
Beddis, R. A., *Britain's Environment: Conserve or Destroy*, Hodder and Stroughton, 1976
Bugler, J., *Polluting Britain*, Pelican, 1972
Dasmann, R. F., *The Conservation Alternative*, John Wiley, 1975
Duffey, E., *The Conservation of Nature*, Collins, 1970
Greswell, P., *Environment: An Alphabetical Handbook*, John Murray, 1971
Hodges, L., *Environmental Pollution*, Holt Rinehart and Winston, 1973
Jackson, O., *Conservation and Pollution*, Batsford, 1970
Lecomber, R., *Economic Growth versus the Environment*, Macmillan, 1975
McKnight, A. D., Marstrand, P. K., and Sinclair, T. C. (Eds), *Environmental Pollution Control: Technical, Economic and Legal Aspects*, Allen and Unwin, 1974
Mellanby, K., *The Biology of Pollution: Institute of Biology Study No. 38*, Arnold, 1972
Mischan, E. J., *The Costs of Economic Growth*, Pelican, 1971
Revelle, C. S., and Revelle, P. L., *Sourcebook of the Environment: The Scientific Perspective*, Houghton Mifflin, 1974
Wallis, H. F., *The New Battle of Britain*, Charles Knight, 1973
Walter, I., *International Economics of Pollution*, Macmillan, 1975
Ward, B. and Dubos, R., *Only One Earth*, Penguin, 1972
Whittaker, C., Brown, P., and Monahan, J., *The Handbook of Environmental Powers*, Architectural Press, 1970
Yapp, W. B. *Production, Pollution, Protection*, Wykeham Publications, 1972.

Atmosphere

Central Office of Information, *Towards Cleaner Air*, HMSO, 1973
Clean Air Council, *Working Party Report on Industrial Emissions to the Atmosphere*, HMSO, 1973
Corey, R. C. (Ed.), *Principles and Practices of Incineration*, John Wiley, 1969
Department of the Environment, *Pollution Paper No 2. Lead in the Environment and its Significance to Man*, HMSO, 1974
Department of the Environment, *Pollution Paper No 5. Chlorofluorocarbons and their Effect upon Stratospheric Ozone*, HMSO, 1976
Medical Research Council, *The Hazards to Man of Nuclear and Allied Radiations: Second Report*, HMSO, 1960
Stern, A. C., (Ed.) *Air Pollution. Volume 1: Air Pollution and its Effects*; *Volume 2: Sources of Air Pollution and their Control*, Academic Press, 1968–69
Warren Spring Laboratory, *National Survey of Air Pollution in the UK, 1961–71*, HMSO, 1972

Land

Baker, F. C., *Derelict Land: 'The Conference in 1970'*, HMSO, 1970

Edington, J. M., and Edington, M. A., *Ecology and Environmental Planning*, Chapman and Hall, 1977

Edwards, A., and Rogers, A. (Eds), *Agricultural Resources*, Faber and Faber, 1974

Johnson, R., *Farms in Britain*, Macmillan Educational, 1972

Ministry of Housing and Local Government, *Report of Technical Committee on the Disposal of Solid Toxic Wastes*, HMSO, 1970

Rudd, R. L., *Pesticides and the Living Landscape*, Faber and Faber, 1965

Thomas, W. S. G., *Land Use and Abuse*, Macmillan Educational, 1972

Water

Bolton, R. L., and Klein, L., *Sewage Treatment: Basic Principles and Trends*, Butterworth 1971

Department of the Environment, Central Advisory Committee, *The Management and Future of Water in England and Wales*, HMSO, 1971

Department of the Environment, *Report of a Survey of Discharges of Foul Sewage to Coastal Waters in England and Wales*, HMSO, 1973

Department of the Environment, *Pollution in Four Industrial Estuaries*, HMSO, 1973

Department of the Environment, *River Pollution Survey of England and Wales: River Quality and Discharges of Sewage and Industrial Effluent*, HMSO, 1975

Edwards, R. W., *Pollution*, Oxford University Press, 1972

Gill, C. Booker and Soper, *The Wreck of the Torrey Canyon*, David and Charles, 1969

James, G. V., *Water Treatment*, Technical Press, 1971

Southgate, B. A., *Water Pollution and Conservation*, Thunderbird Enterprises Ltd, 1969

Warren, C. E., *Biology and Water Pollution Control*, W. B. Sanders, 1971

Warren Spring Laboratory, *Oil Pollution of the Sea Shore*, HMSO, 1972

Noise

Burns, W., *Noise and Man*, John Murray, 1973

Burns, W. and Robinson, D. W., *Hearing and Noise in Industry*, HMSO, 1970

Department of the Environment, *Calculation of Road Traffic Noise*, HMSO, 1975

Noise Advisory Council, *Aircraft Noise Around Heathrow Airport: Flight Routing near Airports*, HMSO, 1971

Noise Advisory Council, *Bothered by Noise? How the Law can Help You*, NA Council, 1975

Office of the Minister for Science, *Committee on the Problem of Noise, Final Report (Wilson Committee)*, HMSO, 1963

Parkinson, P. H., *Acoustics, Noise and Buildings*, Faber and Faber, 1979

Rhodda, M., *Noise and Society*, Oliver and Boyd, 1967

Index of Figures and Tables

Note. Figure numbers are roman; table numbers are italic.

General Index

For key to abbreviations see page 275.

282